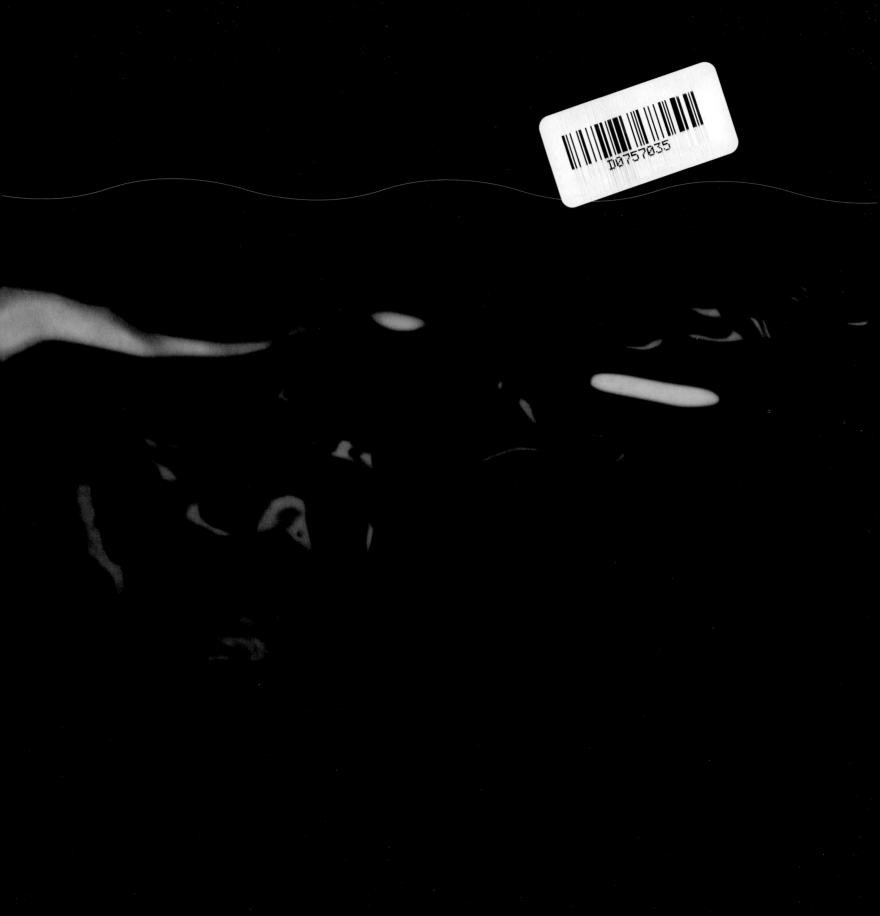

D0757035

Blade of Light
The Story of London's Millennium Bridge

The Penguin Press
in association with
The Millennium Bridge Trust
London 2001

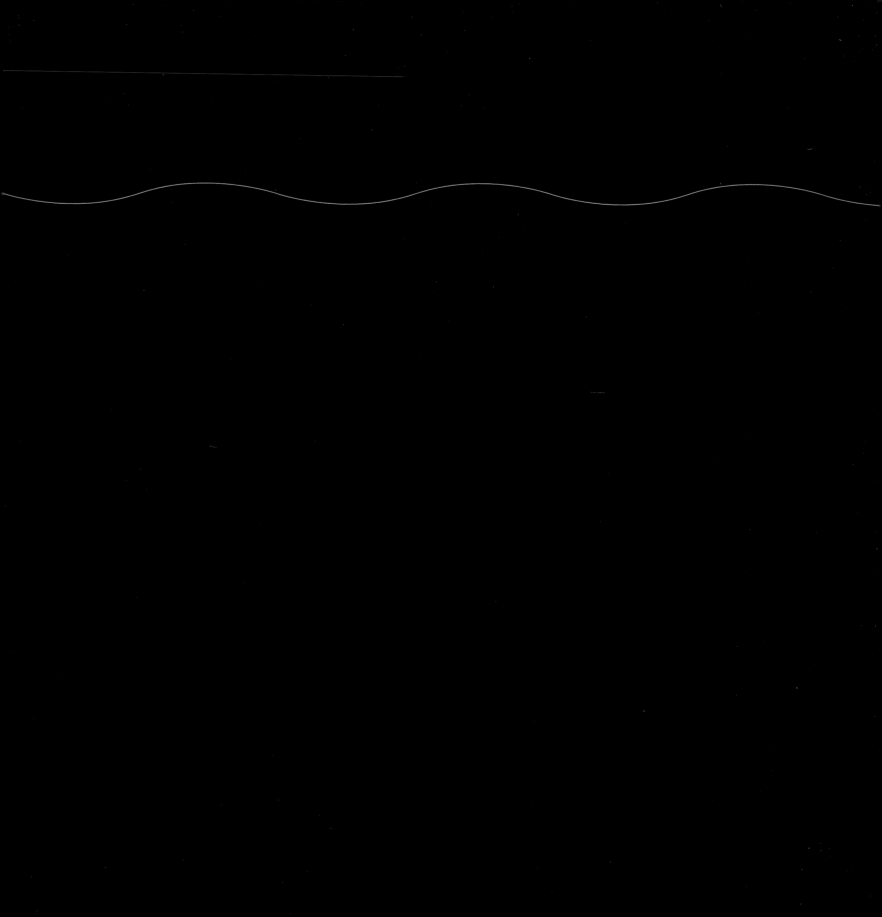

Penguin Books

Penguin Books Ltd
80 Strand, London WC2R 0RL
England

Penguin Putnam Inc.
375 Hudson Street, New York
New York 10014, USA

Penguin Books Australia Ltd
250 Camberwell Road, Camberwell,
Victoria 3124, Australia

Penguin Books Canada Ltd
10 Alcorn Avenue, Toronto
Ontario, Canada M4V 3B2

Penguin Books India (P) Ltd
11 Community Centre, Panchsheel
Park, New Delhi, 110 017 India

Penguin Books (NZ) Ltd
Cnr Rosedale and Airborne Roads,
Albany, Auckland, New Zealand

Penguin Books (South Africa)
(Pty) Ltd, 24 Sturdee Avenue,
Rosebank 2196, South Africa

Penguin Books Ltd
Registered Offices: Harmondsworth
Middlesex, England

First published by Penguin Books
in association with the Millennium
Bridge Trust 2001

This book was made possible by a
generous grant from Pearson plc.

Edited and produced by
Robert Violette, Violette Editions,
London, in association with
Carole Patey

Designed by Atelier Works

Research by Helen Jones, with
thanks to Katy Harris and Elizabeth
Walker (Foster & Partners),
Daniel Imade and Catherine Flack
(Ove Arup & Partners), Nigel Young,
Grant Smith, Dennis Gilbert and
Julian Anderson

Texts by Judith Mayhew, Malcolm
Reading and Savas Sivetidis are
edited extracts from interviews
by Robert Violette, January 2000

Editorial assistance by
Vanessa Mitchell, Chris Patey
and Susanna Scouller

Printed and bound in Italy by
Grafiche Milani

A CIP record for this book is
available from the British Library

ISBN 0-140290-37-0

Contents

6   **Foreword**
David Bell, *The Millennium Bridge Trust*

8   **The Story of London's Millennium Bridge**
Deyan Sudjic

40 **Archaeology**
Robin Wroe-Brown, *Museum of London*

46 **Planning and Funding**
48 Savas Sivetidis, *Cross River Partnership*
52 Judith Mayhew, *City of London*
54 Malcolm Reading, *Millennium Bridge Trust*

60 **Design and Engineering**
62 Norman Foster, *Foster and Partners*
76 Roger Ridsdill Smith, *Ove Arup and Partners*
88 David Newland, *Cambridge University*

94 **Construction**
97 Julian Anderson, *Photographer*

138 **Views**
140 Dennis Gilbert, *Photographer*

150 **Trustees and Patrons of the Millennium Bridge**

152 **Millennium Bridge Project Team**

Foreword
David Bell, Chairman of the Millennium Bridge Trust

The Millennium Bridge is designed to last well into the next century. Long before then it will be taken for granted: as much a symbol of its time as Tower Bridge a century before.

Long before then, too, the circumstances of its construction – and the year that followed its tumultuous opening – will have been forgotten. But this book is about how central London's first new river crossing for a hundred years came to be built and about how and why it became for a time the most famous new bridge in the world. The sudden swaying, 'the wobble', that led to its closure spawned a host of jokes and sparked a fierce argument among engineers which still continues.

But London has taken the bridge to its heart and this book is a celebration of the inspiration, the creativity and the commitment of a host of people over six years. Some are very well known like Ove Arup and Partners, Lord Foster, Sir Anthony Caro and their teams. Others – the concrete pourers, the crane drivers, the whole unsung team of craftsmen – are as 'invisible' as the two massive concrete plugs on which the bridge sits and which were tunnelled 30 metres deep into the river bed. From the start it was a truly international project: the Anglo-Danish construction team of Sir Robert McAlpine and Monberg & Thorsen orchestrated subcontractors from ten countries, from Poland to Norway, Switzerland to Denmark. And Cleveland Bridge, which has since modified the bridge, has been working with equipment from Germany and the United States. To everyone involved we owe much thanks.

From the outset the bridge depended critically on political support from both sides of the river – from Jeremy Fraser, Niall Duffy and Stephanie Elsy, the three leaders of Southwark Council over the last five years, and from Michael Cassidy and Judith Mayhew, who have led the City of London Corporation over the same period. No one could have been more supportive. Sir Nicholas Serota and his team at Tate Modern have been equally helpful. Their new gallery is surely one of the greatest artistic achievements in London in the last 100 years.

Two things came to our aid at critical moments early on. The first was the help that Rab Bennetts and the RIBA gave us in making our original idea of a worldwide competition credible. The second was the Millennium Commission whose early willingness to finance half the construction cost silenced many doubters. Here we are especially grateful to Simon Jenkins. I owe much of my own love of London to the marvellous articles he wrote in the *Evening Standard* twenty or so years ago. His belief in the bridge never wavered.

The project started as a series of partnerships – between a very rich local authority and a very poor one, between English and Danish contractors, between architect, sculptor and engineer. But it would not exist at all but for the further partnership between public and private funds. HSBC gave a fantastic £3 million even before we had planning permission. Sir William Purves and Mary Jo Jacobi, who also chaired our Millennium Bridge Club, were marvellously supportive as has been Sir John Bond the present Chairman of the bank. The Bridge House Estates Trust has been enormously generous and will assume responsibility for the bridge in perpetuity. The Cross River Partnership and English Partnerships have also been very supportive. London Electricity has paid for the lighting of the bridge, Thames Water helped with the Club and Canon gave us the web cameras for our web site, which generated a spectacular four million 'hits' during construction.

David Sainsbury and Hugh Stevenson (who has also been the Trust's Treasurer) gave sizeable personal donations just at the moment when our fund raising risked becoming becalmed. And three hundred people joined the Millennium Bridge Club giving us much-needed extra funds.

We are also indebted to several charitable trusts, livery companies and education authorities in the City and in Southwark, who have supported our education and community programme and to Drivers Jonas, who sponsored the artist, Ron Sandford, to record the building of the bridge from December 1998 to its completion.

Then there is the team that has kept the show on the road over the past six years. Vicki Harvey Piper, then at the *FT*, helped me hugely at the start and is still involved, as, of course, is Julie Sherman my invaluable assistant for the past twelve years. Colin Amery, then the *FT*'s architecture critic, was enthusiastic right from the start and Erica Bolton has given valuable PR advice throughout the project. Carole Patey, the Trust's Secretary, has simply been indispensable as a constant source of ideas and enthusiasm, a brilliant organiser and an incredibly persuasive fundraiser as well. I am grateful too to Eric Gabriel, who with his vast experience of project management gave us wise counsel at a number of crucial stages.

Southwark Council generously took over legal responsibility for the bridge as construction began and had to struggle with the aftermath of its temporary closure .We will be forever in their debt. Bob Coomber, the council's Chief Executive, was unfailingly helpful through thick and thin. Savas Sivetidis and the planning and legal departments were also towers of strength. Malcolm Reading, the Trust's Project Director, and Craig Bradley, Project Manager, have been magnificent.

As an organisation without a 'home' we have been very grateful to Peter Kyle and his colleagues at the Globe and to Ted Hatley, Bursar of the City of London School for Boys (and also a Trustee) for the generosity with which they allowed us to use their buildings for meetings, dinners, receptions and press conferences during the course of construction and for the opening celebrations.

In the City, Andrew Colvin, the Comptroller and City Solicitor, Bill Row, the City Engineer and his team have also given us massive support on the north side of the river. Through it all we have relied heavily on the support and advice of all the Trustees and their names are listed on page 150.

We are delighted that Deyan Sudjic agreed to take on the authorship of this book. He has written a brilliant text that places the building of our new bridge in a wider historical perspective while creating a lasting record of what has been achieved. We are also indebted to Robert Violette who, as editor has been indefatigable in gathering together the wealth of other material.

Finally, I would like to thank Dennis Stevenson and Marjorie Scardino at Pearson for allowing me to devote more time than they, or I, ever imagined would be needed; my colleagues at the *FT* for their unfailing support; and, above all, my wife Primrose, for her enthusiasm and total commitment which sustained me every step of the way.

June 2001

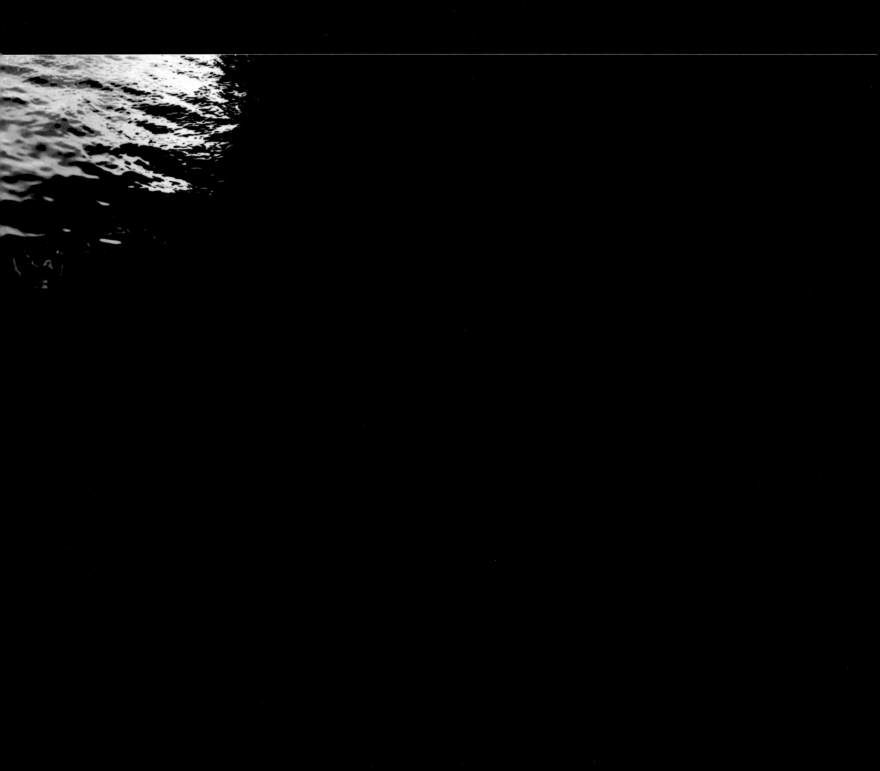

The Story of London's Millennium Bridge
Deyan Sudjic

The Story of London's Millennium Bridge
Deyan Sudjic

As it unwinds quietly downstream past St Paul's Cathedral under a billowing grey sky, with the unmistakable gamey tang of low-tide London mud in the air, the Thames already feels more like the estuary that it is about to become than like a domesticated urban river such as the Seine. Certainly the river is broad enough at this point – a fifth of a mile across. It has a sweep that brings a sense of wide-open spaces right into the heart of London, and a scale that reduces the view looking north from Bankside to a series of smudgy washed-out horizontals, like the marks left by a receding tide on a beach.

By gentle stages, the City steps up from the water to the low ridge that was first built on by the Romans sixty years after the birth of Christ. Despite occasional intrusions from the twentieth century, such as the instantly recognizable saw-toothed silhouettes of the Barbican's apartment towers of the 1960s and the stainless steel rolling-pin that was once the National Westminster Bank's high-rise headquarters, only the dome of the cathedral itself stands out, clearly defined against the skyline. Perhaps that's just as well. Too many of the more recent buildings on the embankment do not bear close scrutiny. Yet the basic pattern of London has survived with remarkable tenacity.

Southwark's foreshore is one of the few vantage points from which it is possible to understand that pattern, and to get a sense of the City of London as a physical entity. You are far enough away not to be caught up in the tangle of ancient streets and alleys, or to have your perspective defined by individual buildings. But you are close enough to get an idea of the city's form. You can see the mark left on London's topography by what, beneath all the reinforced concrete and plate glass, is evidently still the same marshy flood-plain pock-marked with gravel banks that Claudius's legions encountered in the early years of the first millennium.

As they set about the subjugation of Britain, the Romans recognized that they needed a Thames crossing as close to the sea as possible. They found a ford not far from Bankside, and probably very close to the existing London Bridge. That ford soon became the site of the first fixed London crossing, and turned out to be so strategically well placed that it has sparked the growth of one of the world's great cities. From here you can read London's history, economic and political, inscribed in the physical form of the city, as in the rings of a tree. There are still faint traces of the network of roads that is the chief legacy of the Roman city. Half-close your eyes and you can imagine that the 14-foot-high wall that once ringed the city is still standing. In the foreground is the embankment, the most recent in a series of river walls built over the centuries to form the Thames into an artery for trade, and to reclaim precious land from the water and mud. In the background is Terry Farrell's unmistakable post-modern office-block, Alban Gate, marking the line of the wall north of the City, at the point where the Romans had a garrisoned fortress. The fortress's parade ground later made way successively for a Saxon royal palace, a Wren church, and, more recently, a 1960s telephone exchange, which was itself demolished to make room for an office building designed by Richard Rogers.

Over the centuries, the Thames has grown narrower as both banks have been built up and the southern marshes gradually drained. Succeeding generations have moved the river wall further and further forward into the river mud. This area was once a place where people lived as well as worked. The College of Arms, beneath the southern steps of St Paul's Cathedral, with its seventeenth-century rich red rubbed brick facade, formerly the home of the Earls of Derby, is a reminder of what the houses of the rich were once like. The City became a more and more specialized place in the nineteenth century: its homes gave way to the sombre stone offices built by banks and insurance companies, a move which has left a masonry crust that contrasts with the glassy towers of the City's more recent incarnation as the financial centre of the world.

Opposite and above: views of
the site for the Millennium Bridge,
four months into construction.
Foster and Partners / Nigel Young

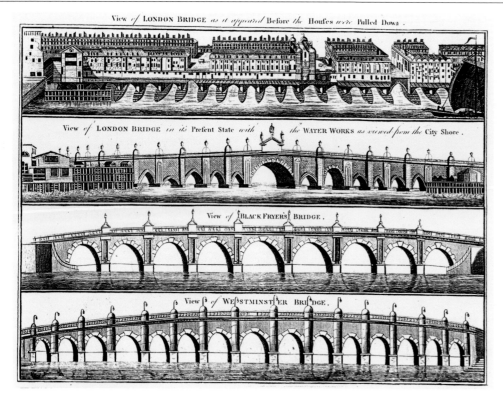

View of LONDON BRIDGE as it appeared Before the Houses were Pulled Down.

View of LONDON BRIDGE in its Prefent State with the WATER WORKS as viewed from the City Shore.

View of BLACK FRYERS BRIDGE.

View of WESTMINSTER BRIDGE.

Anonymous engraving, c. 1760, showing two views of London Bridge, 'Before the Houses were pulled down' and 'in its Present State', and views of 'Black Fryars' and Westminster Bridges.
© *Guildhall Library, London, 2000*

To the right, on the far side of Southwark Bridge, the quietest of all the Thames bridges, is the latest version of the original permanent crossing, London Bridge. In Roman times it was a wooden structure. Sir Nikolaus Pevsner suggests that initially it might even have been formed from a series of boats lashed together, an arrangement depicted in bas-relief on Trajan's column. In the eleventh century, William Rufus, younger son of William the Conqueror, was responsible for the next major step in the history of Thames crossings: he replaced the remains of the old Roman bridge with another wooden structure; he also built Westminster Hall, the core of the Palace of Westminster.

William Rufus can be credited indirectly with the construction of the latest Thames crossing, the pedestrian Millennium Bridge, linking St Paul's with Bankside. In 1097, the Norman king raised a special tax to help repair London Bridge; it was the origin of a fund established to maintain it and three other Thames bridges. Now known as the Bridge House Estates Trust, and under the control of the Corporation of the City of London, that fund has now grown to be worth more than £520 million. It made a donation of £4.5 million towards building costs, without which the Millennium Bridge project would not have been completed.

By the end of the twelfth century, Peter de Colechurch, a cleric turned engineer, had built the first stone London Bridge, which had revenue-generating shops and houses on it. Bridge House was established on the south side of the bridge to administer the funds they yielded. The Bridge House Estates Trust, named after it, now pays for the upkeep of the four eastern-most London bridges administered by the City of London: Tower, London, Blackfriars and Southwark. It will also look after and maintain the new pedestrian bridge. The Trust is a body which must work to a time-scale measured in centuries, rather than decades. It has already established a sinking-fund to pay for the predicted rebuilding of Southwark Bridge in 2073. But the income from the fund is so large that it is able to give away £14m a year to charitable causes throughout London.

Still standing at the foot of Peter's Hill Steps, switch your view away from the north and look towards the south. The picture is very different from that of the clearly defined layers of the north bank. Directly south is the great brick cliff of what is now Tate Modern. Giles Gilbert Scott, architect of the original power station, used art deco with a dash of Aztec. Herzog & de Meuron, architects of its conversion, have left the single chimney, almost as tall as the dome of St Paul's, intact, but have added a giant sculptural beam of light on the roof, visible from passing aircraft as a mark of the building's new use. On the left is Theo Crosby's archaeological reconstruction of the Globe theatre, and on the right is Piers Gough's striking apartment tower for the Manhattan Loft Company. At your back is a remarkable view of St Paul's, at the top of what seems like a river of stone steps, a view that was the creation of the post-war rebuilding of the city, and one of its few unquestioned triumphs. When Lord Holford was planning the precinct around St Paul's, Gordon Cullen was sketching out the creation of a narrow slot, cut into the city's fabric, focused on the cathedral. The planners envisaged it as a place to sit and look out over the river, and then to climb gradually up to St Paul's, and enjoy a view of it on the way as if you were still in the city that Wren knew.

On the south bank, the land is flatter and the urban grain harder to read. Southwark was the community that grew beyond London's defensive wall, one that flourished outside the guild system, but which also ran all the risks that came with the insecurity of being beyond the pale. As in so many cities, it was the south bank of the river that was forced to play the part of poor relation to the north. This former marshland was where London came to take pleasures that were regarded as scarcely licit in the capital proper. It was where, before being shut down by the puritans, Elizabethan theatre flourished – a legacy witnessed by the reconstruction of Shakespeare's Globe just downstream of the Tate. More recently it was a convenient place to put the utilities vital for the continued survival of the city. Workshops, warehouses, wharves and eventually a power station were all built here.

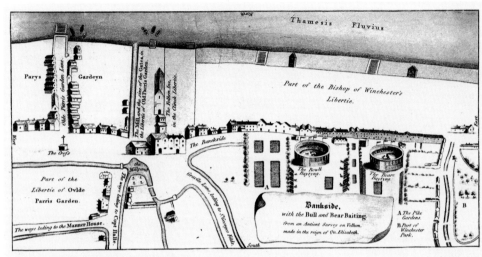

Anonymous Elizabethan engraving, c. 1570, of Bankside, London Borough of Southwark, locating bull and bear baiting rings (centre right) near the current sites of the Globe Theatre and Tate Modern.
© *Guildhall Library, London, 2000*

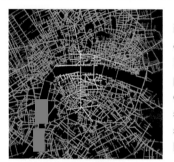

Illustrations by Space Syntax,
University College London,
showing the current distribution
of pedestrian traffic in the St Paul's
and Bankside areas (top) and the
projected increase in pedestrian
traffic after construction of the
Millennium Bridge (bottom).
Foster and Partners / Space Syntax

On the north bank, directly under the gaze of St Paul's, it would have been unthinkable. Land values would have ruled out a power station, even if questions of seemliness did not. But when post-war London went about building a major new power station, the south bank was deemed entirely suitable even as late as 1955. And never mind the fact that the power station presented a daunting and impenetrable barrier to the south that served only to cut off still further the communities living in Southwark's hinterland.

In recent years, nowhere in London has changed more quickly than Southwark's riverfront. Forty years ago it was still a thriving port city waterfront, its wharves thick with barges and river traffic. The container ship killed all that off. And in the last twenty years it has gone from a derelict backwater, characterized by empty warehouses and abandoned factories, to one of London's most vibrant areas. At first offices began to move into the area, attracted by rents much lower than in the square mile. Then it became the focus for some of London's most fashionable residential buildings. And now Tate Modern, perhaps Europe's most exciting new cultural institution to be created in a decade, has established itself here. The area is valued not just for being close to, but not part of the City, but for its own qualities. Now the construction of the Millennium Bridge, running from what was once the site of the pier at which a stream of oil barges tied up to feed Bankside power station's constantly hungry furnaces, to the foot of Peter's Hill, where the stone to build St Paul's was unloaded under Sir Christopher Wren's careful eye, will serve to bring two of London's most disparate areas closer together. They are physically closer than they have ever been: St Paul's is now just a seven-minute walk away from Bankside.

The Bridge will also have an important impact on the psychological geography of London, in the sense that it will change the way that London works and feels. The north bank has some of the most valuable property in the world. It is a pre-eminent global centre for footloose international capital, a vital staging post in the career of every ambitious banker, broker and financial analyst from Zurich to Shanghai. Southwark's hinterland, invisible from the north bank, and until recently hardly visited by those on the north, has been characterized by declining employment opportunities and run-down housing. There are offices and major new employers here, not least the *Financial Times*; but it is still a place where you are as likely to find small businesses operating from lock-up garages and railway arches. It has traditionally been an introverted and close-knit community. The new bridge will accelerate and consolidate the process of rapprochement between the two sides of the river that has been under way for a decade or more. North and south London will never assume the same character, but they will operate more and more as part of a single city, inextricably linked, rather than utterly ignoring each other.

The Millennium Bridge is the first entirely new crossing of the Thames in central London for more than a century: the first new crossing to be built since Tower Bridge opened in 1894. The immediate driving force behind it was the decision by the Tate Gallery to construct a new home for its modern collection in the redundant Bankside power station. The director, Sir Nicholas Serota, settled for adapting Bankside, after having carefully examined a range of options, from expanding on the Tate's original Millbank site (too crowded) to moving to the Docklands (too far away). The power station, not completed until the 1960s, despite the strong flavour of the 1930s in Giles Gilbert Scott's architectural shell, is now home to the Tate's collections of modern art. Its transformation was very much a part of Southwark's plans for bringing its river frontage to life after a long period of neglect; and Serota always saw the possibility of a pedestrian link as a part of the development.

Such a link would be a means of encouraging tourists to extend a trip to St Paul's to include Bankside. Yet the Millennium Bridge is far from having been conceived as simply for the benefit of the Tate. It is part of a much wider urban strategy uniting Southwark's objectives with the broader metropolitan perspective. Southwark councillors wanted to see the business generated by tourism more equally balanced between the two sides of the river, rather than concentrated on the north. The scheme is endorsed and part-funded by the Cross River Partnership, the group bringing together local authorities and other interested parties on both sides of the Thames. The London Borough of Southwark ran the contract for the bridge, in association with the Millennium Bridge Trust set up by David Bell. The completed bridge will come under the auspices of the Corporation of the City of London.

Interestingly, this was far from being the first time that a bridge across the Thames at St Paul's had been mooted. It is an idea that has recurred since at least 1851, when it was proposed by a City councillor. In 1911, the notion got as far as becoming an Act of Parliament, at that time the only legal mechanism by which a crossing of the Thames could be authorised. It took another three years for the Bridge House Estates Trust to organize a competition to design a bridge. One of the reasons for the delay, identified by Felix Barker and Ralph Hyde in their book *London as it Might Have Been* (1982), was doubt about exactly where the bridge should go. One faction perceived the project as a chance to create a more monumental London, with the bridge providing an excuse to carve something resembling one of Hausmann's Parisian boulevards through the City of London. Aligning the bridge directly on the dome of St Paul's and continuing its approach road up to the doorway at the centre of the south transept would enhance the setting of Wren's masterpiece.

The City and the newly established London County Council took a more pragmatic view. They wanted a crossing that would improve north-south communications, with the bridge at St Paul's providing the basis of a link from Aldersgate to Southwark. The LCC believed a bridge would be useful for a new tram route from the Angel to the Elephant & Castle, but only if it was positioned to hit the north bank to the east of the cathedral, rather than aligning itself on the dome. In 1914, with entries from a number of distinguished architects, including Albert Richardson, who later designed Bracken House, the old *Financial Times* headquarters, the competition was won by G. Washington Browne, who submitted a grandiloquent classical design, marked by twin sets of sentinel pylons at either end, and a winged goddess driving a two-horse chariot. At this point the First World War intervened. Despite an attempted resurrection of the project in the 1920s, the idea of a bridge at St Paul's was ignored by Patrick Abercrombie when he drew up the LCC's plan for the post-war reconstruction of London.

In the 1990s, perhaps intrigued by the ambiguous relationship between St Paul's and the Bankside power station, with its chimney almost as high as the dome of the cathedral, more and more architects and engineers began to look at the possibility of designing a bridge here. Santiago Calatrava, architect of the groundbreaking 1992 Alamillo Bridge in Seville, exhibited a proposal for such a bridge at the 1996 Royal Academy of Arts exhibition *Living Bridges: The Inhabited Bridge, Past, Present and Future.* The charitable Peabody Housing Trust marked the centenary of its founder George Peabody in 1995 by inviting six architects, including John Outram Associates and Allies & Morrison, to think about how the site might be used to create an inhabited bridge incorporating a range of speculative uses, including housing, a university and a new building for St Bartholomew's Hospital.

The Millennium Bridge was from the start considered on a much more modest scale, and was a practical rather than a theoretical proposal. It was conceived as exclusively for pedestrians, not merely in terms of allowing them to move from north to south and south to north, but also providing them with a view. Of all the vantage points to explore London, the centre of the river is perhaps the most privileged, and the Millennium Bridge is the only crossing on which the din of traffic or the rumbling of railway locomotives does not disturb the pedestrian. But as with earlier attempts to bridge the Thames at this point, there was still an issue about exactly where the crossing should go, its precise relationship with St Paul's, and also with the reconstructed power station, with its strongly symmetrical facade.

The creation of the bridge is the product of the combined efforts of a great many people: the teams of specialist contractors who have had to work in often difficult conditions, in coffer-dams deep in waterlogged Thames mud, to pour deep concrete pile foundations; the component manufacturers from as far away as Poland and Finland; the designers, who have organized, uniquely, a working partnership between artist, engineers and architects; Southwark Council; the Corporation of the City of London; the Millennium Lottery Fund; private sponsors, including the most substantial contribution from HSBC. If there has been one individual without whom the project would never have happened, it is David Bell, of the Pearson media group.

In his earlier incarnation as managing director of Pearson's subsidiary, the *Financial Times*, Bell occupied an office at the newspaper, hard by the southern flank of Southwark Bridge and looking directly out over the Thames. Turning to the left at his window, he could see St Paul's and the tower of the old power station. It was then that Bell began to think about the possibility of putting the resources of the paper behind a proposal for a new bridge. 'We were very keen to help contribute something for the millennium that belongs to everyone, which lasts and which is free.'

A new pedestrian bridge certainly met those criteria. The *FT* has always had a connection with architectural excellence. It has a long tradition of distinguished architectural correspondents, going back to Howard Brockman. It has supported an award for industrial architecture, and has itself been responsible for commissioning such distinguished buildings as its old HQ, Bracken House, and its former printing works, which was designed by Nicholas Grimshaw. What seemed to be required was a project that would distil architectural creativity at the turn of the century. To quote Bell again: 'The *FT* had the opportunity, and also a responsibility, to leave something behind for London.' Bell's personal inclination was towards something architectural. 'I started my career as a journalist on the *Oxford Mail*, and by default, I was, among other things, the architecture correspondent.' All things considered, it seemed appropriate that the paper should assume a leading role in helping to realise the idea of a new and inspirational pedestrian bridge.

In 1996 Bell went to the Royal Institute of British Architects (RIBA) for advice on the project, and they suggested an architectural competition. It was a step that would follow in the tradition of competitions for Thames bridges; Tower Bridge, for example, was the product of one. A competition was duly arranged and, with 227 entries, revealed just how much professional interest there was in the design of a contemporary bridge. By the mid-1990s, bridge design had experienced a radical turnaround. The 1970s were years in which it had seemed that the highest objective of the bridge engineer was to design a structure that was as inconspicuous as possible: to deny, visually, that anything dramatic was involved in spanning water.

From the mid-1980s, however, a series of spectacular designs showed the extent to which a bridge could be a powerfully expressive sculptural object. Bridge designers around the world began to look for ways to push forward the limits of what was possible, with each new project inspiring the next. And in many ways, the Millennium Bridge, with its remarkable and prominent site, provided a chance for the summation of their efforts.

The competition went ahead before it was clear how the bridge across the Thames would actually be funded, or even exactly where it would be located. In retrospect, Bell, who has spent the last six years working on the project, unpaid, as and when his managerial responsibilities at Pearson allow, is surprised as much by how complex the process of turning the winning design into a physical bridge has been as by how much has been achieved in such a comparatively short time. 'Six years ago we had the first meeting with all the interested parties we could identify. "You will never do it," they said. "You will never get planning permission; too many people don't want this to happen." I had never done anything like this before. I had a suspicion it might be somewhat difficult, but I had not expected that there were thirty-six different permissions to get. It is a high profile place to do anything. But I didn't think it was going to be quite so complicated as it has turned out to be. You solve one problem, and another one emerges. The cycling lobby, for example, objected because you can't cycle across; they wanted an extra lane. But that would have gone against the concept of the bridge. On the other hand, the ecology class at the City of London School for Boys, who are right by the northern end of the bridge, didn't like our original plan to use teak for the bridge deck, even though it would have been plantation grown. So we switched to a more sustainable material, aluminium.'

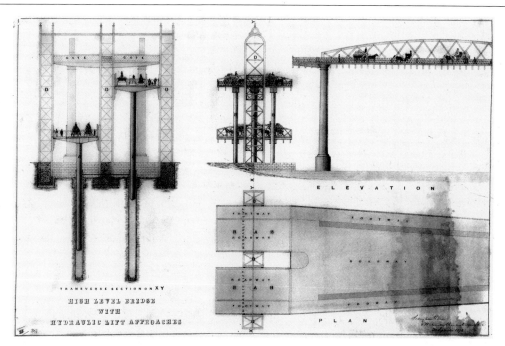

Design for high-level Tower Bridge, 1872, by Sidengham Duer. Hydraulic lifts carry horses, carts and carriages to the bridge 80ft above the river.
© *Corporation of London*

Looking at the conditions outlined in the brief for the competition, it is remarkable just how onerous and demanding all the approvals were, to say nothing of just how many wildly disparate interests had to be consulted and satisfied. The conditions begin with the straightforward 'The design life should be 120 years' and 'The bridge must be capable of withstanding the impact of a 3000-tonne displacement vessel travelling down river at 12 knots'. Clear enough, but then things start to get complicated. Any new bridge must, for example, meet the provisions asked for by a body known as the Environment Agency Thames Region under the Land Drainage Bylaws (1981) and the Thames Region Flood Defence Bylaws (1991). Then competitors were warned that 'Proposals for dewatering on the north bank will require the approval of the dean and chapter of St Paul's', a body not normally known for their land engineering expertise, and must in addition 'comply with the St Paul's Preservation Act (1935)'.

Perhaps the most bizarre restriction of all was the need to keep out of the way of a post office tunnel under the Thames, whose exact location must be kept secret 'for security reasons'. Competitors could not be trusted with its exact wherabouts, but were just told that 'it is assumed to be within the area under review. Contact with the post office will be required for approval.' To the experts, however, the most demanding provision of all was the Environment Agency's insistence on 'an hydraulic study of the effects of the bridge design on the river regime'. On my photocopied copy of the conditions, an unknown engineer has ringed the word 'hydraulic' and scribbled 'Bastard!' He at once recognized a request that would entail elaborate and time-consuming large-scale testing in water tanks.

Bell, however, is not easily deterred. He is a believer in architecture. But his other passion is for the achievements of the great Victorian engineers, and in particular, for those of Sir Joseph Bazalgette. 'He is my real hero. He dealt with nineteenth-century London's sewage problems by building four great steam-driven pumping engines to get rid of the Great Stink. They pushed the sewage out of London down the river and into the open sea. They were just too big to demolish and they are still there to this day.' To Bell, the new bridge is a contemporary echo of Bazalgette's combination of daring engineering and public spirit. 'Those pumping stations were leading-edge technology in their day, just as the new bridge is at the edge of what is possible now. And they did a lot to make London a better place to live. The bridge is an opportunity and a responsibility, one that has got massive emotional appeal.'

Inevitably it was not a view immediately shared by everyone. While the Tate and the London Borough of Southwark were keen to see a new link from the City to the south bank of the river, there were some who were concerned that nothing should be done that might interfere with the magnificent views of St Paul's. They were sceptical about the need for the bridge in the first place. And these objections came on top of an endless list of practical problems, from how to raise the necessary funds to questions about the legal position on bridge-building in central London. Would it require a special Act of Parliament, as had been the case for the last four bridges to span the Thames in London? Or was it an issue (as it eventually turned out to be) that could be resolved under the conventional planning system? In the event, the Port of London Authority used its statutory powers to grant a licence for a river crossing, and the City and Southwark both granted planning permission for those parts of the development within their respective jurisdictions. Then there was the question of the ownership of the land needed to build the bridge, and even of precisely who would be ready to take on the leadership role needed to push for the resolution of all these complex issues, and have the patience, diplomacy and persistence to deal with them all.

Statement by the Millennium Bridge Competition Jury

A response to questions raised by competitors during the first stage of the *Financial Times* / Royal Institute of British Architects architectural competition:

The Competition Jury
David Bell Chairman, *Financial Times*
Ana Patricia Botin President, Banco Santander de Negocios
Michael Cassidy Chairman, Policy and Resources Committee, Corporation of London
Anna Ford Broadcaster
Councillor Jeremy Fraser Partner, Southwark Council
Jacques Herzog Architect, Herzog & de Meuron, Zurich
Sir Phillip Powell Architect
Frank Newby Consulting Engineer
Sir Michael Perry CBE Chairman, Unilever
David Sainsbury Chairman, Sainsbury's
Wilfried Wang Architect and Director, Deutsches Architekturmuseum Frankfurt

Q.1 With regard to the existing loading jetty in the river adjacent to Bankside power Station:
a) Is it still in use in any way and is it needed?
b) Are there any proposals to repair and improve it?
c) Can it be removed as part of the bridge scheme?
A.1 a) The jetty is currently not in use and there are no known operational reasons to retain it. It may be of use during construction of the bridge and/or the Tate Gallery of Modern Art (TGMA).
b) There are no proposals to repair it or improve it. The Tate Modern has an option to purchase the lease, the owner being, The Port of London Authority. It may be assumed that the basic structure is sound. The competitors may propose uses for it.
c) It may be assumed that it could be demolished, but the cost of this needs to be taken into account. An advantage of demolition would be to open up views of St Paul's from the South Bank which are currently obstructed by the jetty.

Q.2 No budget has been mentioned. Shall we assume that the sky is the limit?
A.2 Value for money must be assumed to be a significant factor in the competition.

Q.3 Please elaborate on the meaning of plan ME3 [provided in the competition brief].
A.3 Plan ME3 shows the area zones within which any proposals will be judged against the individual policies. For further explanation of the policies, see paragraphs 6.17– 6.20 and Appendices A and B [of the competition brief].

Q.4 Is the existing Bankside jetty to be retained?
A.4 See Question 1.

Q.5 Under '5.0 Prizes' [in the competition brief], you refer to 'each second-stage competitor'. There is no mention of any 'second stage' in the document. Please clarify.
A.5 The competition is being held in two stages. Those competitors selected to proceed to the second phase will each receive an honorarium of £5 000. At the onset of the second stage, a second-stage briefing document will be issued to the shortlisted competitors.

Q.6 With reference to points 5.8 and 6.14, please clarify if 'pedestrian only' means that no cycles are to be allowed onto the bridge.
A.6 Current by-laws exclude cyclists from the footpath on the north bank of the river between Southwark and Blackfriars bridges and from Peter's Hill. For the competition, it should be assumed that the bridge will be pedestrian only.

Q.7 Is there any reason why the general pavement level of the riverside walk can't be the same as the Flood Defence Level of 5.41?
A.7 In considering the level of the walkway, account needs to be taken of flood defence requirements, access to and amenities of adjoining properties, safety regulations and maintaining the integrity of the existing [river] walls.

Q.8 What is the exact site boundary of the TGMA
A.8 The land ownership in the area may well be different by the time the bridge is built. Both the London Borough of Southwark and the Tate Gallery wish to see the land between the TGMA and the river designed as a single entity integrating the bridge, and therefore the current proposals shown on plan E126-0-011 may well be revised to reflect the final bridge solution. The TGMA entry point shown on the north side of the building must be safeguarded.

Q.9 Are there any specific existing constraints under these terms that need to be taken into account if making a proposal in this area?
A.9 A car park lies under the central and western part of the terms but will not be operational. This should not be considered as a constraint.

Q.10 Can the existing power station landing stage be demolished and the connecting tunnels be demolished of infilled?
A.10 See Question 1 also. The large oil and access tunnel is to be filled as part of the TGMA works and the other tunnels will be plugged at both ends. The small tunnels may be filled if required by the design of the bridge.

Q.11 Is there a budget?
A.11 See Question 2.

Q.12 Is a cost plan expected with submission?
A.12 No, not as part of the first-stage submission.

Q.13 Are cycles to be provided for or excluded?
A.13 See Question 6.

Q.14 Under 10.3 'skid resistance' [in the competition brief], what is DB.29/87?
A.14 This should read BD 29/87 Department of Transport. Design Criteria for Footbridges.

Q.15 Is it possible to illustrate examples of previous work on submission boards or would this compromise anonymity?
A.15 Examples of previous work should not be included on the A2 [-size] submission boards as this could compromise anonymity.

Q.16 Can the area contained by Peter's Hill be considered as a potentially treated area?
A.16 The inclusion of this area will be subject to meeting the various constraints and issues set out in the brief, as set out, in particular, in Section 6.0, and must also respect owners and occupiers of adjacent properties. Public access on walkways and views along the vista must be respected.

Q.17 Can the area around the front of Bankside also be potentially treated?
A.17 See Question 8 also. The objective for the competition is the design of the new bridge, but it is recognized that competitors may wish to address the area immediately adjacent to and relating to the new bridge.

Q.18 Are there any constraints associated with the existing Bankside landing stage or may it potentially be removed?
A.18 See Question 1.

Q.19 On [section] 6.18 [of the brief], 'There should be no… etc' (last sentence): Does this mean no design under any circumstances will be allowed if it infringes St Paul's Heights in any way? Does this apply to the north half of the river only or does it also include the limitation projected to the south bank as shown on fig. 7?
A.19 The design and engineering solution for the proposed bridge must ensure that the structure will maintain local and strategic views of St Paul's Cathedral, and competitors should refer to Policy ENV28 of the City of London UDP in Appendix A. Figure 7 shows the extent of the St Paul's height limitations, which is to the centre line of the river, but the effect on views of a development on the south side of this centre line will be an important consideration.

Q.20 On 6.19 'views from the monument, etc': What does 'impact on' mean?
A.20 In respect of impact on views from the Monument, competitors should refer to Policy ENV30 of the City of London UDP in Appendix A and Plan ME3. See Question 3 also.

Q.21 'Alternative design under separate cover…': Do these need new individual pink declaration forms and entry fees? Or will a photocopy of the form suffice?
Q.22 Anyone wishing to make more than one submission for the competition should pay an additional registration fee to the RIBA [Royal Institute of British Architects] in order that a new declaration form can be issued and a record made of the new registration.

Competition entry by Georg Rotne and Niels Gimsing, Copenhagen.
Millennium Bridge Trust

It was Bell who took on this task, building relationships with politicians and officials on both sides of the river, and commissioning research to see just how many people would use a bridge if one were built. MORI interviewed 460 people working in all parts of the city: two-thirds of them supported the project, and three in ten said they would use the bridge once a week, with more than half saying they would use it at least once a month. Bell also began to engage in fund-raising, and took the preliminary steps to submit an application for financial support to the Millennium Commission. Above all, he set in motion the process that would actually produce a design, perhaps the most effective response to those who expressed concern about the impact of an unsympathetic design on such a sensitive setting. After discussions with Colin Amery, at that time the *Financial Times's* architecture critic, Bell announced that the paper would fund an international competition, organised by the Royal Institute of British Architects, to find the best possible design for the bridge. The brief for the competition, drawn up in the summer of 1996, set out to encourage collaboration between artists, architects and engineers. The brief specified that this would be a bridge of 'exceptional quality in design and engineering, a world class structure that will in itself be a landmark and a visitor destination'.

Bell recruited a high-powered jury for the competition, whose eleven members included Frank Newby, the distinguished engineer responsible for the aviary at London Zoo, and Jacques Herzog, of the Swiss architectural practice Herzog & de Meuron, who were carrying out the transformation of the former Bankside power station into the new Tate Modern. Also on the jury were Michael Cassidy, chairman of the policy and resources committee of the Corporation of London, and Jeremy Fraser, Southwark Council's leader, along with Anna Ford, David Sainsbury, the architect Sir Phillip Powell and others. With technical help, they sifted through the 227 submissions, which were presented and judged entirely anonymously.

A short list of six teams was drawn up, and those six were asked to work up their schemes to a more finished level. The schemes themselves were still anonymous, but with the shortlisted teams named publicly, there was of course a great deal of more or less informed speculation about who was responsible for what.

Two weeks before Christmas in 1996, David Bell called a press conference in the City of London School for Boys to introduce the winning team of Norman Foster, Anthony Caro and Ove Arup and Partners and their design, previously referred to only as scheme number 166. 'This is a very exciting moment,' he said. 'The Millennium Bridge will be London's first new Thames crossing since Tower Bridge opened in 1894. It will be London's first bridge for pedestrians only, drawing together the traditional trading centre of the nation in the City of London and a burgeoning new cultural quarter on Southwark's Bankside.' The submissions showed an enormous range of approaches. The technical assessors endorsed all six in terms of buildability, and their chances of being realized on budget. The assessors were free to make a choice based on the qualities of the designs that they saw. The competition brief had left a considerable degree of freedom – even the question of exactly where the bridge would touch or, in some cases, not touch the bank, was left open. The clearest constraint was that this would be a bridge for pedestrians only, and that it would be special enough to live up to its exceptional setting.

Georg Rotne and Niels Gimsing, the Danish team with a long track record which included working on the design of the epic crossing of the Oresund between Denmark and Sweden, proposed a bridge slung between a pair of cables supported on high level piers. Frank O. Gehry, architect of the remarkable Bilbao Guggenheim, working with Richard Serra, developed a swelling, organically shaped, asymmetric bridge, with a profile like an elongated soup spoon. Also shortlisted were Cezary Bednarski, with Studio E Architects and artist Peter Fink, McDowell and Benedetti and the Tokyo-based architects Ushida Findlay. The engineers Dewhurst Macfarlane featured in two of the submissions.

Competition entry by Frank O. Gehry
Associates, Santa Monica, California,
with artist Richard Serra.
Millennium Bridge Trust /
David Lambert

Competition entry by Ushida
Findlay Partnership with Dewhurst
Macfarlane & Partners, London.
Millennium Bridge Trust /
Marcus Leith, David Lambert

**Competition entry by Cezary
Bednarski/Studio E Architects,
London, with artist Peter Fink.**
*Millennium Bridge Trust /
David Lambert*

Competition entry by McDowell
& Benedetti, London.
Millennium Bridge Trust /
David Lambert

Scheme 166, competition
entry by Foster and Partners and
Ove Arup and Partners, London,
with artist Sir Anthony Caro.
*Millennium Bridge Trust /
Marcus Leith*

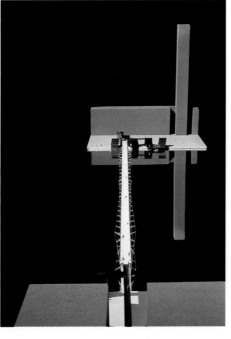

Scheme 166
Millennium Bridge Competition
Foster and Partners, Ove Arup and Partners, Sir Anthony Caro

Extracts from a document prepared in November 1996 by Foster and Partners, Ove Arup and Partners and Sir Anthony Caro as part of the design team's revised scheme for the Millennium Bridge at the second, shortlist, stage of the architectural competition, on the basis of which Scheme 166 was announced as the competition winner on 10 December 1996. Some specifications and detailed design of the bridge – such as those for the south terminus, the ramps and steps on the north and south sites, the handrails and the material for the decking, which is now aluminium instead of wood – subsequently changed during development of the bridge scheme in 1998 and 1999.

Introduction

This bridge is the ultimate expression of spanning the greatest distance with the minimum apparent means. There is an aesthetic elegance in this expression of performance and economy which is in a long-standing tradition of design. By day the bridge will be an extraordinarily thin 'blade' of stainless steel and cable, whilst at night it will appear as a 'blade of light'. This apparent dematerialization of the bridge enhances the experience for the pedestrian of views both from it and through it.

But the movement of people is also vertical as well as horizontal – so the bridge itself describes an elegant arc which touches down lightly in the passage of Peter's Hill and terminates grandly on Bankside. All vertical connections are by gentle ramps – ideal for pedestrians and wheelchairs alike – with shortcuts by small flights of steps which interlock with the ramps. At Bankside, these create a curved ziggurat, which integrates bridge, sculpture, building and viewing platform. At its heart there is a café or shop, which can be entered from the upper bridge level or the riverside walk over which it hovers. There is also a version that is not inhabited, with a void at its core – but the social focus of a living bridge with refreshment or shopping is a compelling idea in this location. It would integrate commercial pressures rather than these becoming random events in front of Bankside. Both versions are

The lowest part of Peter's Hill, between the point of contact of the bridge and the riverside walk, is reconstructed as a cascade of ramps and steps, all in Portland stone. The bridge is less than a third of the width of the passage, so that views of St Paul's are not interrupted. Moving towards the cathedral, the junction of Queen Victoria Street is marked by two small-scale ziggurats or stepped gateways. These signal the link with Bankside from afar. At night the route is traced by sunken lights in the pavement between the 'gates' and the bridge itself.

The 'feel' of the bridge is tactile and sensory. The edge trim, lighting grills, handrails and balustrades are of stainless steel, surfaces which are enhanced by use and wear, not defiled. The colour is integral with the material and will improve with the imprints of age and usage. Stainless steel, with a finish which is not too shiny, also picks up the colour of the day, reflecting the quality of light which depends on the weather and the time of the year. … The structure within the bridge is galvanized steel, exposed only on the underside where it forms a delicate lattice. …

Is this bridge architecture, engineering or sculpture? It has been largely created by three individuals, from these three individuals – a sculptor, [an] engineer and [an] architect. As authors of the project, we see it as a shared creative act. The opportunity to further develop our first-stage proposals has led to a much greater creative integration and this is reflected in a bridge which is itself now totally integrated. For example, the first proposal showed a crossing which was level from one side to the other. By changing this to a gentle arc, it is not only more graceful but slimmer. There is also a significant benefit in access. The curved section lowers the end of the bridge at Bankside by nearly two metres, resulting in a saving of more than 40 metres of ramp. The integration of ramps and steps into the urban fabric of Peter's Hill has simplified the connection of the bridge to the north side; this is a very significant advance on our first proposals.

On the south side, we debated all the options for this critical landing. Our original scheme sought to show how this space might be colonized by growing out from the point of contact with the bank and extending the sphere of influence of the crossing. But it is frankly not possible to second-guess the true intentions of those who are responsible for the area in front of the new Tate and we believe that it would be presumptuous to lead with designs for this critically important civic space. We have therefore sought to achieve a balance between strongly asserting the presence of our bridge but with a deference to the wider urban context, which should ideally be led by the designers of the new Tate.

PLAN

ELEVATION

Engineering and design philosophy

Description of Structural Elements
The structural depth of the footbridge is relatively shallow because the self-weight is light, as are the loads imposed by pedestrian traffic. The potential problems arising from the static and particularly the dynamic behaviour of the bridge have been thoroughly checked. Two high-strength steel cable tendons flank the lightly curved deck. From their highest point over the piles at the centre of the main span, the cables dip three metres in elevation and curve two meters towards the centre of the bridge on plan.

The steel deck structure is fixed to the cables every eight metres by tapering hollow sections. A rectilinear grid of tubular sections provides a rigid structure both in torsion and in plan, whilst also acting as the secondary support structure for the deck surface.

The weight of the bridge is supported by cables which are anchored on the north and south banks by groups of tension and compression piles inclined at 14° to the vertical. The resulting pre-tension in the cables is such that the horizontal forces imposed on the river piers are restricted to those arising from imposed loads. As a result, piling in the river is minimized and the piers are relatively slender. They are in reinforced concrete, elliptical in shape and topped out by a fabricated steel bracket which reaches out in a 'V' shape from under the deck to pick up the steel cable. The helical ramp on the south side is an independent structure on piles comprising a series of beams underneath the ramp which radiate out from the centre. …

Resistance of Structure to Static and Dynamic Actions
The bridge must resist static and dynamic actions principally due to river traffic, pedestrians and wind. As well as being strong, the structure must satisfy serviceability criteria, which limit deflection and vibration of the bridge. The primary support for the bridge against static loads is provided by the cables acting in tension on each side of the deck. Examination of the structure showed that, dynamically, the bridge responded as a beam of varying section, with the deck edge beam and the cables acting as chords, and the tapering arms between them transferring the shear. This system was shown to be far stiffer than that due to the pre-tensioned cables alone.

River traffic is prevented from contact with the piers by 'dolphins', which act as buffers. The piers need therefore not resist the loads due to a collision and can retain a lighter and more elegant and economical form.

Pedestrian traffic exerts static imposed loads either over the whole deck or in symmetric or asymmetric patterns. The relatively low resultant tension due to these loads in the cables, compared to the initial pre-tension, means that the live load deflections are acceptably small.

Wind exerts both static and dynamic forces on the bridge
The rigidity of the deck structure, coupled with the horizontal curve of the low cables, means that the bridge is very stiff in plan. Static wind loads are resisted by the combination of these two systems and result in minimal deflections.

Wind Tunnel Test
In order to evaluate the aerodynamic stability of the proposed bridge, a series of wind tunnel tests were performed on a sectional model of the central span. These tests were carried out by the Canadian company Rowan, Williams, Davies & Irwin (RWDI) [Guelph, Ontario]. The bridge's properties were modelled on a 1:16 scale section of the structure and the dynamic response measured using a strain gauged suspension system in the 2.4-metre wide by 2-metre-high boundary layer wind tunnel.

The wind tunnel tests were conducted to assess the potential for vortex-induced oscillations of the deck and to determine the wind speed at which the bridge deck goes into divergent oscillations (i.e., flutter or galloping) for three angles of attack (-3°, 0° and +3°). The bridge section was examined at three different positions along the central span:
Cable Position 1: base deck with cables at level of deck edge beam centreline (approximately 24 metres from the piers);
Cable Position 2: base deck with cables below deck edge beam centreline corresponding to a point approximately 32 metres from the piers;
Cable Position 3: base deck with cables below deck edge beam centreline corresponding to the mid-span point of the central span.

Finally, the deck was tested without any cables present as a reference case to assess the aerodynamic stability of the deck. Conservatively, the bridge was assumed to have a torsional structural damping ratio of 0.5% of critical and to have a balustrade with 50% porosity. No vortex-induced oscillations were noted during any of the tests. No galloping oscillations (self-excited vertical vibrations) were observed in the tests. Flutter – self-excited aerodynamic instability, which may involve torsional motion only or coupled torsional and vertical motion – is typically the most important effect to assess for a bridge of this type. However, a combination of the tubular deck members and the inclined cable geometry in section produces a torsionally stiff 'U' section with a relatively large mass polar moment of inertia. In addition, the fanning out of the cables enables them to act as aerodynamic damping elements on each side of the deck.

Tests demonstrated that the critical flutter velocity of the bridge was over 45 metres per second (160 km/h) for all cable positions more than twice the 50-year mean hourly windspeed for London. The mean windspeed is comfortably exceeded even without any cables on the bridge. RWDI stated that the tests assumed not only conservative bridge properties but also a smooth wind flow towards the structure. In reality, the wind in a built-up area, such as this, would be very turbulent and therefore improve stability. Variations in the angle of attack of the wind showed that the damping capability of the cables could be further improved by slightly inclining them to the horizontal. They concluded that the bridge would be stable in wind speeds of up to 50 metres per second (180 km/h) and that stability could be even further improved through adjustment of the cables and refinement of the balustrade design.

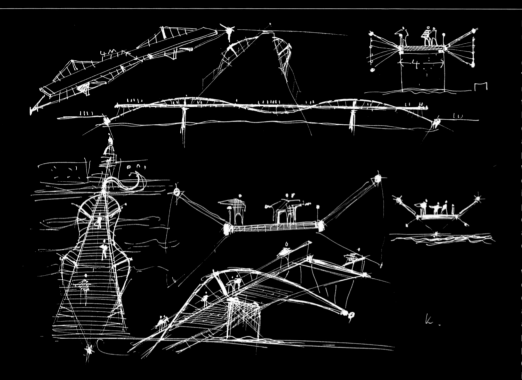

Preliminary construction method statement

General Considerations

The construction methods used are all in accordance with the Construction Design and Management (CDM) regulations of 1994. They have been developed, along with the sequence and programme timing, in consultation with Schal, one of the leading construction managers in the industry, who foresaw no special problems in building the bridge, other than those normally associated with construction over water.

In order to simplify construction which must necessarily take place over tidal water in regular use and on banks with restricted access, all the major structural elements are composed of prefabricated parts, including the pilecaps, the piers, the cable tendons and the deck. Interference with river traffic is minimized. All river work will take place from a temporary anchored barge. Apart from this, the [navigation] channel will remain entirely open, even during the placement of the cable tendons.

Lighting strategy

The initial concept of a 'blade of light' is retained and enhanced. The soffit of the primary structure will be lit along the length of the bridge, creating a thin 'blade of light' across the river. The light source will produce a crisp white light, to accentuate the qualities of the three main materials. Two continuous light sources illuminate the edge of the walkways. These will also uplight the piers. The walkway will be spotlit to make the walking surface clearly visible for the visually impaired. This rhythm of lights is then continued by spotlights set flush in the ground up to Queen Victoria Street. The columns in the river will be gently washed with crisp white spotlights. Two orange lights, one metre apart, will be installed to highlight the navigation channels, with a floodlight coming on as boats move under the new bridge.

The whole bridge could also be covered by a closed-circuit television system linked into the security network for Bankside, at eight-metre centres, the primary structural grid.

Access for all

Our proposals are designed to achieve full access for every member of the public. The slope on the arc of the bridge is considerably less than 1:20, as are the slopes of the helical ramp on the south side. This ramp turns through 180° on several occasions, creating many opportunities to rest. The ramp is over two metres wide, making it easy to pass.

On the north bank, we recommend that the steps be reorganized, with a combined ramp/step arrangement for everyone. Ramps at 1:20 would have landings every eight metres and will be 1.5 metres wide between handrails.

All surfaces will be slip-resistant. Throughout the scheme we intend to use two balustrades. The higher one is 1200 millimetres high and 250 millimetres wide, to allow people to lean against it to enjoy the view on both sides. The lower balustrade is 900 millimetres high and 65 millimetres in diametre to enhance children's sense of security on the bridge and to be readily available for anyone who needs to use a handrail. The handrails are set at specific heights to miss the eye-level of wheelchair users.

The panels below the balustrades are curved to make it very difficult for people to climb onto the tension rods in the handrails. The handrails are the same timber as the deck and will be dark so that they can be clearly seen in contrast to the stainless steel balustrades. Information posts with Braille texts in different languages would be provided at certain locations on the bridge and on the viewing platforms. These would describe the view, and also have induction loops to offer the hard of hearing information.

The use of the bridge by bicycles would require careful consideration. We would recommend that cyclists are asked to dismount and push their bicycles across the bridge to avoid any conflict with pedestrians. Access for all is a vital ingredient for the success of this scheme, to be developed in consultation with all the relevant authorities.

Description of materials and maintenance

... Galvanised and Stainless Steel

We would propose using galvanised steel for all primary structural elements and stainless steel everywhere else where it is visible (i.e., on the ramps and balustrades.) Steel production is a relatively high-energy consuming process. However, because of its strength, it allows the quantity to be minimized in terms of the amount of material used.

Galvanising is the most successful and simple solution to corrosion protection. Stainless steel not only offers the advantage of being less energy and material demanding in its total life-cycle analysis, being made of scrap metal, it also has the great advantage of being corrosion protected and independent of chemical protection.

Materials and corrosion protection

Deck structure and 'V'-shaped plate girders on piers

Material: Grade 50 steel.
Corrosion protection: Blast-cleaned, then painted with Zinc-rich epoxy primer, two-pack epoxy micaceous iron oxide barrier and acrylic urethane undercoat and finish.

Cable tendons

Material: high-strength steel parallel strand.
Corrosion protection: each stand individually galvanised, ensemble-coated with wax and sealed in polypropylene skin, 1.5 millimetres thick. ...

Piers

Material: reinforced concrete.
Corrosion protection: exposed finish, with sufficient cover between reinforcement bars and exterior surface to avoid corrosion.

Maintenance Programme

The bridge should be regularly inspected so as to maintain it in a good state of repair. The painting specification is designed for 15 years, after which the structure must be thoroughly inspected and

Speech by Lord Foster given at the press conference to announce the winning design for the Millennium Bridge Competition, 10 December 1996

'It is a very great pleasure for me to be standing here with Tony Caro and Chris Wise as part of the team chosen for this wonderful project. I really do want to emphasize that collaboration is of the essence here. To work with Tony Caro, and with Chris Wise and his team at Ove Arup and Partners, is, for the Foster partnership, a true privilege.

'To have the opportunity to work on a venture of this calibre and profile in London is an honour indeed. A new pedestrian bridge for London is a scheme close to our hearts, and has, I feel, exactly the significance that a millennial landmark should have. It will link two very different sectors of a great city. It will further the regeneration of Bankside and will enhance the life of the City of London and all London. Above all, it will be an expression of optimism, and of regard for both Londoners and visitors – a bridge for people.

'By coincidence, Foster and Partners is currently involved with the design of two bridges in Europe: a giant viaduct for a motorway in a spectacular part of the Tarn valley in southern France, and a railway bridge crossing the wooded straits of Arstaviken into Stockholm. But London's Millennium Bridge is, for us, very special. To have a chance to build something in London – where our only two completed projects are the ITN Headquarters and the Sackler Galleries at the Royal Academy of Arts – is a wonderful challenge. I would like to thank the judges of the competition, and all of you, very much.'

Scheme 166 won because it was the most minimal, most elegant of all the submissions: so slender and so attenuated as to hardly register on the skyline, and yet seen from the roof-top restaurant of Tate Modern it is a dramatic and striking sculptural object that blurs the distinctions between design and architecture, and between sculpture and engineering. Sir Anthony Caro, one of Britain's most distinguished living sculptors, played a key role in the winning submission. Interestingly enough, he had studied engineering at Cambridge before shifting direction and going to work as an assistant to Henry Moore. After establishing his own studio, he moved from stone to increasingly large-scale steel assemblies that often used bright colour. He began to work with architects on public installations of his sculpture, perhaps most famously with I. M. Pei at the extension to the National Gallery in Washington. In such an arrangement there are likely to be frustrations: 'Pei worked with me by saying "There is a gap here, please will you fill it." But for a sculptor, that is not being really involved with architecture, it's adding to architecture.'

Sculpture and architecture have had a long relationship, stretching back to classical times, when one was an integral, physical part of the other. Sculpture escaped from this role in the nineteenth century, shrugging off the plinth, and moving on to the table and the ground in the last century. But it has yet to satisfactorily renegotiate the terms of its sometimes uncomfortable relationship with architecture. The issue has exercised Caro over many years. He has worked on pieces that have begun to approach the scale of architecture, and in addition has developed a keen interest in architectural practice. Thus he was intrigued to learn of a competition that insisted on architects, engineers and artists working together from the earliest stages of the design. His previous collaborative experience had led him to expect last-minute encounters at which the sculptor might be asked to do nothing more than add a decorative detail to a virtually complete project.

It was Caro who called Norman Foster to suggest a collaborative effort for the competition: 'I gave Norman a ring, and said, "What do you think?" He was very positive: "Great idea. Let's meet." So I made a few sketches beforehand. I thought about a giant table piece from one side of the river to the other. Norman came to my studio with Ken Shuttleworth, one of his partners. He also brought Chris Wise, a dynamo of an engineer from Ove Arup.' Roger Ridsdill Smith, a young engineer who had studied at Cambridge, like Caro, and who joined Ove Arup after a spell in Paris, was also at the meeting: 'Chris Wise saw the competition announcement. At Arup we thought about doing it ourselves, but we had been coming second in too many bridge competitions. Then Norman asked Chris if we would help, and it seemed like a better strategy.' That first meeting took place in October 1996. Caro's studio is an old Victorian factory, very much a working artist's studio, with an extensive workshop and a team of assistants. Over a well-scrubbed long table, the bridge rapidly took shape. 'Everybody came with little ideas, but there wasn't an actual scheme,' Ridsdill Smith remembers. 'Chris and I had a chat in an Italian restaurant beforehand. We talked about arches, and suspension systems, but we couldn't get a handle on it. Then Chris took a napkin and drew two banks and a straight line between them. There was no structural idea, just the thought of a long, straight, thin blade. Horizontal bridges don't look good when people stand on them; they tend to sag, so we had to find a way to deal with that. The other idea was to address the remarkable location in mid river. Some suspension bridges are just a corridor of structure, which means that you don't look around much as you cross them. The cables on each side mean that your view is directed. We wanted a platform, a flying carpet as thin as possible – below, as well as above the deck.' As Caro remembers it, Wise's suggestion galvanized the meeting. 'Chris came in and said, "I see this as a guitar string pulling the two banks together." It was such a grand thought, so convincing. I know enough about how things stand up, but I didn't know that you could make a suspension bridge that sticks out to the sides, rather than above.'

From this point, the centre of gravity for the design meetings shifted from Caro's studio to Foster and Partners' offices overlooking the Thames at Battersea. 'It was Norman who brought things together,' says Caro. It was a useful place to work. The meetings were held around a large circular table in the corner of a room from which there is a striking view of Albert Bridge. Albert Bridge is a light suspension bridge, which both inspires and warns would-be bridge designers, its elegant lines fatally compromised by a makeshift additional support at centre-span, propping up the structure.

Foster himself played a leading role in the design process, assembling a small team to work on the competition entry and eventually to see the bridge through to completion. Ken Shuttleworth was the partner in charge, Andy Bow was the Project Director, and the Project Architect was Catherine Ramsden, a recent recruit who had joined the firm just three weeks earlier. Bow's most vivid memory of the early meetings: 'Tony [Caro] and Norman had wonderful conversations about how to make the bridge land on the south side. We used anything that came to hand around the table to try to express their ideas three dimensionally. On one occasion we were piling up brown sugar cubes and got the office model-makers to come across with bits of elastic.' The eventual submission was the culmination of an intense period of work. As Bow recalls: 'Just before the hand-in, Norman himself came down to the print room, and started colouring in the presentation drawings by hand.'

Top: Ken Shuttleworth, Lord Foster and Sir Anthony Caro meeting at Foster and Partners' Battersea offices, London, in autumn 1996, to discuss Caro's design for the south end terminus of the bridge during the first stage of the architectural competition. Above: brown sugar cubes and elastic bands were used by Foster and Caro in one meeting to model the bridge's proposed span and abutments.
Foster and Partners / Rudi Meisel / Visum

The authors of Scheme 166 did a great deal of preparatory homework. Even at the first stage, when the competition brief asked for nothing more elaborate than a couple of conceptual sketches and 500 words of description, the project team had set up a series of wind tunnel tests in a specialist Canadian laboratory. So determined were they to prove that the structure would not suffer from the traditional problem facing suspension bridges, the dynamic wind-induced flutter that famously brought down the Tacoma Narrows Bridge in 1940 just four months after it was opened, that they invested in a 1:16 scale model of the bridge and a great deal of computer time. But what they did not predict was a previously little understood phenomenon that would cause the bridge to sway when filled with pedestrians, a problem that would cause the bridge to close immediately after its official opening.

The competition submission for the second stage became a more conventional architectural and engineering solution than anything that Caro would have submitted on his own. Nonetheless, Caro felt closely involved. 'People ask me which part of the bridge is yours,' says Caro. 'And of course I can't point to any single aspect of it. But that isn't really the issue. The idea of a sculptor, an architect and an engineer working together is what really makes this bridge different, and I was there when all the important decisions were made. Asking a sculptor to do a handrail at the last moment is the old-fashioned way of working. We did it the new-fashioned way, and that was wonderful. I've learnt that architects think differently from us artists. We can say things they wouldn't dare to, like "Why not try it upside down?" We think from an unexpected standpoint. We don't take practicality into account. On the other hand, they can tell us quite a bit about scale: we were beginning to make things 100 feet high. I could see the process as pointing the way towards a new way of working, and even a new way to educate artists. Why aren't architecture schools next door to sculpture schools, for example? You can talk to a painter when you are at art school, but why can't you talk to an architect?' But Caro acknowledges that the practical constraints of bridge design weren't all to his taste. 'I don't enjoy working with committees, and I can see what architects and engineers have to go through.'

With the bridge itself moving towards an engineering and architectural solution, Caro began to focus on the points at which the bridge would touch land at either end. 'We made little models that explored a sequence of long ramps, to address the disabled access problem.' In six weeks the concept was distilled into two sheets of A1. The computer visualisations capture the essence of the scheme as it has been built. But there have been considerable modifications to the way in which the bridge meets the south bank of the Thames. As originally conceived it would have taken the form of an unravelling ribbon of steel, stepping back and forth on the edge of the river wall, mixing giant steps with swooping curves, and integrating a sequence of long ramps for access, a complex set of forms which certainly carried Caro's sculptural signature, and which was echoed in more modest form on the north bank. 'It was what got us into the last six', says Caro. The design team then had to respond to the feedback from the judges in the final stage of the competition. 'I've heard that what we have done is regarded as too sculptural,' Foster told one meeting of the team. The response was what Caro calls the 'cake'. Instead of his original notion for the bridge approach on the south bank, the team came up with a more constrained proposal. The bridge itself did not change substantially, but at the point at which it reached the south bank, it was met by a set of tightly spiralling curves, set around a central lift (with a small restaurant also provided), looking out over the river. It was reminiscent of a wedding cake. Caro recalls Ken Shuttleworth attempting to negotiate its tight curves in a wheelchair to test its practicality for himself: 'He kept bashing the side.' The decision was also taken to gently arch the bridge, rather than cross the river all on the same level. 'We got a letter from the judges, the gist of which was that they wanted more integration between architecture, engineering and sculpture,' says Bow. 'We studied that letter for days. We looked at every word, over and over again, and thought about it, and tried to figure it out'.

The design for the south end of the
Millennium Bridge evolved through
several stages; the final design
(bottom right) is a variation on the
'eye of the needle' scheme.
Foster and Partners / Nigel Young

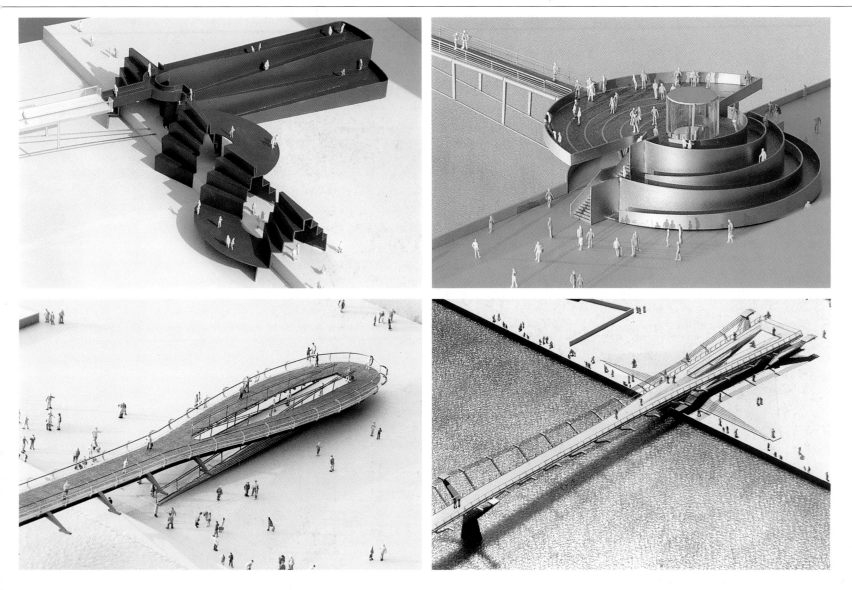

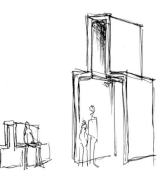

For the second stage of the competition, the design team not only changed the profile of the bridge (in so doing reducing the need for part of the access ramp): it took a different aesthetic tack. For the first stage, the bridge was a horizontal white glass plate that terminated with red oxide, strongly sculptural pieces. In its revised version, instead of a sharp distinction between the bridge and its land-based adjuncts, between slim architectural engineering and expressive sculpture, the new scheme used a timber deck and stainless steel substructure to unify both elements. These changes, and the design team's working methods, were reflected in the report that accompanied the revised design. 'Is this bridge architecture, engineering or sculpture?' it asked. 'It has been largely created by three individuals from the three disciplines – a sculptor, engineer and architect. As authors of the project, we see it as a shared creative act. The opportunity to further develop our first-stage proposals has led to a much greater creative integration and this is reflected in a bridge, which is itself now totally integrated. For example, the first proposal showed a crossing which was level from one side to the other. By changing this to a gentle arc it is not only more graceful but slimmer. There is also a significant benefit in access. The curved section lowers the end of the bridge at Bankside by nearly two metres, resulting in a saving of more than forty metres of ramp.' Though their scheme won the competition, it went through a further modification, a variation of the so-called 'eye of the needle' version, which is what ended up being built. It is a lower-profile solution. And its adoption meant that Caro's main visible contribution is a series of large-scale sculptural pylons to mark the presence of the bridge at the north end.

At the Fitzrovia offices of Ove Arup and Partners, Britain's leading engineering consultancy, another team was set up to work on the bridge. Sir Jack Zunz, one of Arup's most distinguished partners, came out of retirement to assist with the early stages. Chris Wise and Roger Ridsdill Smith led the Arup effort, while Sophie Le Bourva, qualified as both engineer and architect, took on a co-ordinating role, bringing together all the different engineering specialities needed for the project. 'We staged regular meetings in the office, very much like an architect's reviews, but instead of putting up sketches or sample boards, we had the engineering numbers pinned up on the wall', she says.

From the structural point of view, the major constraint on the engineers was the need to keep the height of the suspension structure supports as low as possible so as not to interfere with the view of St Paul's. The solution, a flat suspension bridge consisting of a deck and two piers in the river, came as a logical outcome. It may not look like a suspension bridge: there are none of the characteristic towers to hold up the cables; but that is what it is. Rather than rise above the bridge deck, the cable bunches are grouped on either side of the structure, like the outriggers of a canoe, barely rising above the level of the main bridge structure. Ridsdill Smith calls it 'a flat suspension bridge, one that is particularly long and skinny'. To enable them to take the strain the cables are anchored in massive concrete blocks, one at each side of the river, sunk deep enough to transfer load down beyond the soft infill soil near the surface, and positioned so as not to damage the existing river bank retaining walls.

In the early stages, the competitors were not limited to a defined budget. But even without financial constraints, Ridsdill Smith believes that the team would have come up with a broadly similar strategy. 'This wasn't a case of looking for a wonder material. We might have come up with a solution that suspended the bridge from one bank to the other, with no piers at all. It would have been a striking idea, but would not have made financial sense.' What places the bridge at the limit of what is technically possible are the advances in engineering technique. 'Ten years ago, there was no software to test it. There is much more computer power available now'. But even so, the first calculations were still done by hand. What makes a bridge such a demanding design task is that every aspect is on show, and on show from every direction. It means that every screw fixing, every weld, every bolt connection is considered in detail, not just as a practical functional issue, but also as an aesthetic one.

The other novel aspect of the bridge is the extent and degree of collaboration between engineer, architect and sculptor – three separate entities even before one adds the various manufacturers of components. To achieve this level of integration, Ridsdill Smith and Le Bourva split the management of the project between them, Ridsdill Smith liaising with members of the team outside Arup, while Le Bourva co-ordinated the expert contributions from within the firm. 'The sizes of the cables and fittings were the most important element informing the design', she says. 'We all went to find out how suspension cables are made, and how they are sized. We went to look at the Humber Bridge. There is not only one engineering solution to a given problem, there is a level of subjectivity, and that is what makes it interesting.' Le Bourva's and Ridsdill Smith's own interest in architecture helped to ensure a smooth working relationship between architects and engineers in the team. 'Catherine Ramsden from the Foster office came to all our sessions. It meant that the architects never got sudden decisions without them knowing about it in advance.'

With the Foster/Caro/Arup design named the winner, the challenge moved to securing the necessary planning permissions, a process that involved refining the design, and also a patient process of persuasion. There was a simultaneous effort to raise the necessary funds to build the project. To that end, David Bell worked with Savas Sivetidis, the Southwark officer responsible for co-ordinating the plan and technical brief for the bridge competition. Sivetidis appointed Malcolm Reading as Project Director at the beginning of 1997, submitted an application for a Millennium Commission grant on behalf of Southwark Council, and set about taking the practical steps to establish the Millennium Bridge Trust as a legal entity to raise matching funds. Carole Patey was appointed Appeal Director and Secretary of the Trust in the summer of 1997. In the event, the Cross River Partnership made an initial donation of £1 million towards setting up costs. Thereafter, the Millennium Commission awarded £7.2 million to the project, and matching funds were raised from the Bridge House Estates Trust, from several private donors and from HSBC, the largest corporate sponsor. All of this depended on securing planning consent, and an endless round of presentations to interested amenity groups to win hearts and minds.

Wind tunnel test of the bridge deck and balustrades at laboratories in Guelph, Ontario.
Ove Arup and Partners

Southwark's planning committee was the first to give its consent. Winning approval from the Corporation of the City of London was a more demanding business. Norman Foster himself made the presentation, persuasively describing the concept while a five-minute video portrayed the walk that users would enjoy as they moved across the bridge towards the City, with St Paul's getting larger and larger at each step. This proved an irresistible argument in favour of the bridge design and its position, but there followed a series of detailed negotiations about the shape and form of Caro's sculptural markers, designated as the HSBC Gates, signalling the presence of the bridge on the north bank. Caro is directly responsible for two large pieces of stainless steel that mark the presence of the bridge on Queen Victoria Street, and two smaller pieces that people can sit on: 'The markers are there to say "This is where it is"'. The exact positioning was a matter of considerable experimentation and thought. 'We had a meeting in the studio about it one day, and we all decided to go down to the site to see for ourselves. There were ten of us, and we got people to stand in a line down the steps, and then move down to where the bridge would be. I did a bit of shouting and got them all to stand in a row. It really needs markers all the way up to St Paul's, but the planning committee wouldn't buy that.' As it is, the markers are flat and have been rotated so that St Paul's is framed between them.

Meanwhile, the structural aspects of the bridge were also being explored and tested. The structure had been examined in the first stage of the competition in the world's biggest wind tunnel, at a laboratory specialising in bridge dynamics at Guelph, Ontario. The conclusion was that the bridge would survive all but a 1-in-10000-year gust. The effect of the bridge on the behaviour of the river was also tested, this time using an even larger model, twenty metres long. The worry that the bridge would increase the scouring tendency of the tides proved groundless.

Then there was the need to design the bridge to resist possible collision impact from river traffic. The number of boats passing in a given period were counted to assess the risk, and the shape of the bridge piers was modified to deflect impact. The thinking was that while a bridge can have cables or deck sections replaced, the loss of a pier would mean a fundamental structural failure. The most substantial change at this stage was the switch from a timber deck to an aluminium one. Despite the design team's confidence that they had found teak that could be sourced from a sustainable plantation crop, aluminium was used to avoid controversy. 'In my naivete, I thought it would simply mean getting out the drawings, scoring out every reference to timber, and substituting aluminium,' says Andy Bow. 'In the end, it led to the total redesign of the whole decking system, and so of the bridge itself.' The engineers initially believed that the weight of the teak played a stabilizing role, and considered a means of adding more mass. But in the event, they found that it had no effect, and that the substructure could be made lighter without difficulty.

A full-size mock-up of a section of the bridge deck was made to explore its physical characteristics in detail. This helped to ascertain the correct height for the rail. Claude Engle, responsible for lighting the bridge, which has such a critical impact on its appearance at night, found himself having to advise the architects on how to detail the light fittings in such a way to meet concerns from the City that they would need constant vacuuming to remove the charred remains of electrocuted gnats and moths in the summer months.

In parallel with the development of the design work, the small client team, David Bell, Savas Sivetidis, Malcolm Reading and Carole Patey, was strengthened by the Southwark appointment of Craig Bradley as Project Manager. There followed a period of intense activity to get all the necessary consents and land issues in place and to raise sufficient funding for the project to go ahead. Work finally began in December 1998, when the Museum of London carried out an archaeological dig, four months long, on both the north and south banks. In among the traces of Roman, Saxon and mediaeval defensive walls and embankments, the archaeologists found hollow-log and brick arched culvert drains, a sunken barge, clay tobacco pipes and lead seals used to mark cloth bales.

Construction of the bridge began in April 1999, the work shared by a team of two firms: Monberg & Thorsen (builders of the giant Oresund Bridge in Denmark) and Sir Robert McAlpine. They were already collaborators on the refurbishment of the Forth Bridge. For this new venture, McAlpine took responsibility for the substructure, Monberg & Thorsen for the superstructure. The first step was the construction of coffer-dams, made of elliptical steel sheets, to allow the piles to be driven deep into the river bed. For stability they had to go past the London clay levels. This was not an easy operation, and involved sending teams of miners down to dig it all out, with pumps working continually to keep the shaft dry and a giant fan blowing down air to keep the work force alive. After the tunnelling and the pile boring were complete, the steel reinforcing cage went in. Then came the concrete. In September 1999 there was a ten-hour-long concrete pour to complete the abutment on the south side of the river, involving 1000 cubic metres of concrete, ferried in 160 lorries from three different sites in south London. This mass of concrete covers the sixteen piles to a depth of twenty metres, and carries the full force of the cables.

The north bank was completed shortly afterwards, and the piers finished, so that work could start on the bridge superstructure in October 1999. In January 2000, the giant pier arms were installed, followed by the installation of two temporary A-frame structures which supported a hoist across the river to winch the steel cables into position. The deck and the balustrades, shipped from fabrication sites in mainland Europe and assembled in the Thames estuary, were brought up river in sections.

For the design team that had spent the last three years thinking about how the bridge would look and feel, and relying on models and computer images, it was an important moment. The physical reality was almost there at last. 'What is great about the bridge is that it is a kind of kit of parts, like a Meccano set,' says Catherine Ramsden. 'The components were being constructed all over Europe, and we had been flying around to see all these things in isolation. It gives you a very abstract feeling. But then they started to appear on site, and suddenly the bridge was really there.' It was a far-flung construction process. The suspension cables are from northern England, the cable clamps from Norway. The bridge wings were built in Finland, as were the pier brackets (the outstretched 'arms' that sit on top of the piers), which came down the Thames in the second week of January, to be introduced to the deck-carrying brackets fabricated in Poland. These brackets had been assembled in Rochester and hold the aluminium decking that came from Germany. Every single step of this manufacturing process was critical not just to the structural performance of the bridge but to its architectural quality, and demanded endless site visits. As Andy Bow says: 'You can build all the models you want, but the quality of the product comes down to the person who is welding that actual joint, that actual balustrade, somewhere in the middle of a very cold factory in Denmark or Poland.'

Full-size maquette of a section of bridge decking, inspected on site by the design team, Bankside, September 1999.
Foster and Partners / Nigel Young

H.M. The Queen with the Lord Mayor of London and the Mayor of Southwark at the dedication ceremony for the Bridge, 9 May 2000
Foster and Partners / Nigel Young

In warm sunshine on 9 May 2000 HM The Queen dedicated the bridge. The day began at St Paul's cathedral, packed for a special service. Afterwards the Queen and Prince Philip attended a reception at the City of London School for Boys to meet people who had played significant roles in the planning, design and construction of the bridge, and to view an exhibition of artworks by local schoolchildren. Accompanied by the bishops of London and Southwark , the Lord Mayor of London and the Mayor of Southwark, and by David Bell, Chairman of the Millennium Bridge Trust, the Queen and the Duke processed onto the bridge to the sounds of music on the river and riverside performed by brass ensembles from the Guildhall School of Music and Drama and the City of London School. A specially commissioned fanfare sounded from both sides of the river and Her Majesty formally dedicated the bridge. Beneath, oarsmen from the Thames Traditional Rowing Association rowed by in their long clinker-built boats. A steel band from Oliver Goldsmith's School reminded the crowd of the rich cultural diversity of the capital city. Those fortunate enough to have received invitations then made their way to Guildhall to enjoy the traditional hospitality offered by the City Corporation on great occasions.

A month later a spectacular firework display marked the lighting of the bridge. Two days later, just after 9am on a sunny Saturday, the Millennium Bridge first began to wobble. Roger Ridsdill Smith was waiting for the start of a sponsored walk, one of a weekend of events staged to mark the official opening. As the bridge began to fill with people, he spotted the first sign of movement. 'It happened quite fleetingly, but you could tell it wasn't just a judder; it was a resonant movement. I thought, "that's interesting", but as the numbers thinned out, the bridge calmed down again. Then at lunch time, when the bridge finally opened to the public, the crowds swarmed on from both ends. And that is when the bridge really started to move. All 690 tons of the steel and aluminium deck began to sway left and right, so much so that pedestrians suspended above the Thames on slender steel cables began to clutch at hand rails to steady themselves against the sway, to stay upright. As they did so, the swings began to get increasingly violent. We asked ourselves, "Is it dangerous?" You could see that the movement was self-limiting. Beyond a certain point, it simply becomes impossible to get any more people on the bridge. As it gets fuller and fuller, people stop moving and the effect subsides, long before structural safety limits are reached. We knew it wouldn't fall down, but suppose somebody fell and broke a leg or ankle?' The bridge was closed, while the design team, to the accompaniment of enormous media interest considered what to do next.

'It's not what is different about the bridge that caused the problem', says Tony Fitzpatrick, who led the Arup team that solved the mystery of what causes the sway. 'Every issue about the bridge that was innovative was properly researched, and worked perfectly. What hit us in the back of the head was that bit of the bridge that was the same as every other long span bridge. We assumed that it would work like other bridges have until now; that was the mistake'.

Left: Firework display to celebrate the lighting of the Bridge, 8 June 2000, by Christophe Bertonneau who also designed the Millennium Eve display at the Eiffel Tower, Paris. *Foster and Partners / Nigel Young*

What Arup discovered was, they say, a previously misunderstood phenomenon. Once the number of pedestrians on a bridge passes a critical limit, their footsteps start to make it move, and the more that they react to that movement to stay upright, the more the bridge shakes. It could potentially affect any pedestrian bridge over a given length – up to 70 in the UK alone. Arup carried out urgent checks to see if the bridge had been built exactly as specified – it had – and then they looked yet again at their own calculations to see if there were any elementary mathematical mistakes in the numbers. There weren't.

All bridges are liable to move. Their weight and structure keep them still to a certain extent, but get enough people walking across a bridge and all the natural damping is cancelled out; then the next few footsteps will set it wobbling violently. It's not a gradual effect. Arup found that it is all or nothing. 'Put 1,000 people on a bridge, and it will seem to be fine. Put on 1,100 and it starts to move' says Fitzpatrick.

Arup discovered a number of bridges which had suffered from the problem, and every one of them looked utterly different. There is a high level suspension bridge in Tokyo, completed in the 1980s, that had a makeshift damping system retrofitted after its opening; a 100 year old steel truss bridge in Canada that never moved a millimetre in its entire life, until the day of the firework display on its centenary that attracted so many pedestrians that it began to wobble. Why had none of these cases forewarned Arup? Fitzpatrick's answer is that they were never reported to the people who write the codes that bridge designers must follow.

Arup's solution for eliminating the problem was subjected to a series of in-situ tests before work started on installing it. The solution involves fitting a pair of X-shaped braces under each of the structural bays under the bridge, along with 37 viscous dampers (the kind of large shock absorbers that you might find on a truck), and another 50 tuned mass dampers. These are heavy blocks that sit in baths of oil and are connected by springs to the structure. When the bridge starts to move, the blocks absorb the energy triggered by pedestrians and stop the movement. It's all been done deftly enough for the additions to look like part of the original design.

Now that it is at last complete, London's first new bridge since 1894 is a powerful work on the boundaries of art, architecture and engineering that succeeds in distilling these disciplines down to their essence. There is no room for camouflage, or for a gradual unfolding of ideas, like a novel. A bridge is more like a three-dimensional poem. Everything is visible, nothing is hidden. It is as beautiful as the smallest bolt head, and as the springing curves of the suspension cables, achieving a satisfying balance between the solidity of its graceful piers and the lightness of the aluminium deck. It is a sensitive addition to the cityscape, in two enormously different settings on north and south banks. Beautiful to look at and to look out from, it is an architectural achievement, an engineering triumph. It is a tribute to the people who laboured in Thames mud to build it, to the politicians and officials and funders who made it possible, and, of course, a tribute to the enthusiasts who were determined to build it in the face of all the obstacles and scepticism they encountered. Watching the bridge take shape, stage by stage, reveals how bold a move any new crossing of the Thames really is. It represents a fundamental shift in the nature of central London's geography. It affects the great views of London, and powerfully modifies the urban spaces created by the river. Under the eyes of St Paul's, London now has a remarkably modern new structure, to which the much overused term 'uncompromising' really applies. And yet it already feels as if it has always been there.

Archaeology
Robin Wroe-Brown

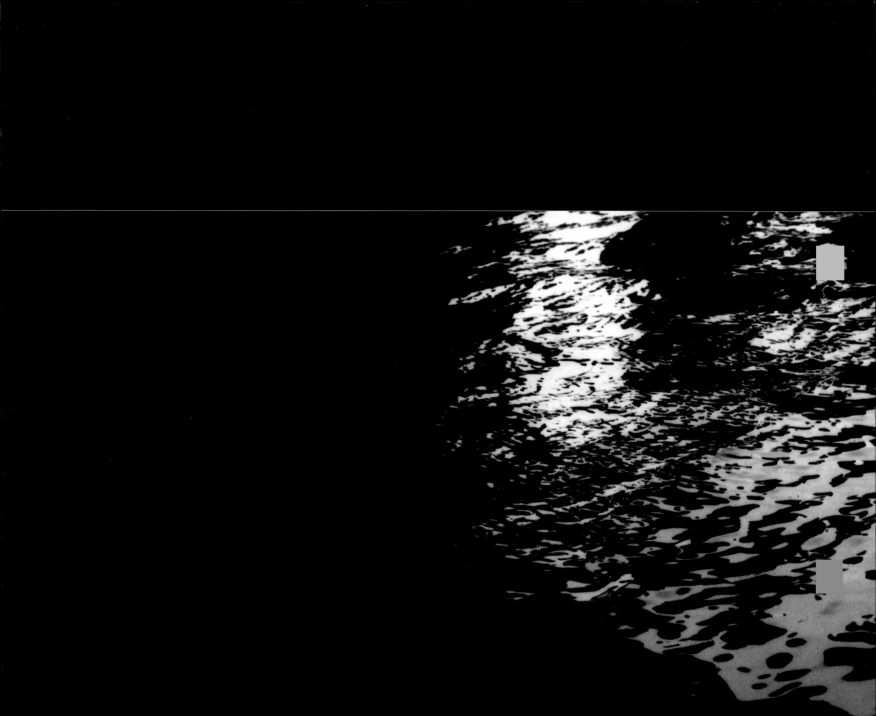

Archaeology
Robin Wroe-Brown, Senior Archaeologist, Museum of London Archaeological Service

Below, from top to bottom: the excavation on the north site in progress, with Boss Alley revealed between the parallel walls in the foreground; an eighteenth-century brick river wall exposed on the Bankside excavation; detailed drawings were made by MoLAS surveyors of this inlet wall on the north site. *Maggie Cox, MoLAS*

In recent years archaeology has become an integral part of almost all major construction projects in London. The construction work for the Millennium Bridge has presented archaeologists with an exceptional opportunity to examine the ancient waterfronts on both sides of the river Thames. Full-scale excavation began on the north bank of the river in December 1998 and on the south bank in February 1999, conducted by an experienced team from the Museum of London Archaeology Service (MoLAS). This continued through March and the same team was transferred to the north site at the beginning of April to maintain continuity. The final trench was formally closed on 28 May. The outcome certainly lived up to our expectations.

The Thames in London during Roman times was a great deal wider: Southwark was an expanse of marshland and channels and the City river edge was some 100 metres further north than at present. Since then the margins of the river have been narrowed by reclamation and consolidation over the centuries, with each successive advance covering the previous with large dumps of earth. As a result the timber structures, called revetments, built to retain the dumps, were often simply buried. Dockside activity was conducted on top of the newly reclaimed land and buildings followed behind the advances.

Archaeology in a waterlogged environment tends to exhibit many distinct characteristics. Because the soil here remains waterlogged, the timbers frequently survive in excellent condition, displaying complex carpentry and toolmarks. Crucially, the timbers can often be dated with extreme accuracy, occasionally to a particular season in a year, using tree-ring dating. Other organic remains survive equally well and much can be learnt about the environment of the Thames from plant and insect remains. Artefacts such as leather, textile and metal objects also emerge in very good condition.

As expected, the discoveries on the south bank ranged in date from about the fourteenth century to the present day, with the southern half of the site producing the earliest material. A succession of timber revetments were recorded in varying states of survival. The earliest had been almost entirely dismantled in antiquity and the timbers were probably employed in the construction of the next. A brick wall formed the waterfront of the seventeenth and eighteenth centuries and in front of the wall a further three timber river walls were identified. The final wall on site was constructed in the 1930s from concrete and metal sheeting.

Behind the revetments on the west side of the site a remarkable sequence of drains was excavated. The deepest (and therefore earliest) was a hollowed log, followed by a square drain built with planks, a brick and timber drain and an arched brick culvert. All of these drains emptied into the Thames at the place which became Mason's Stairs, an inlet in the river bank on the east side of the site. Unfortunately much of the evidence for the inlet was destroyed by modern basements and drainage.

A thirteenth-century decorative
jug handle.
Andy Chopping, MoLAS

Above, from left to right:
Ornate copper alloy horse harness
pendant from the medieval period;
an elaborate copper alloy dress
hook, with open-work decoration,
of the late fifteenth or sixteenth
century; two views of the lead
head for a fourteenth-century
stick puppet, from which a cap of
distinctive form has been removed
(a more complete example can
be found elsewhere in London);
part of a thirteenth- or fourteenth-
century copper alloy strap-end with
a stylised depiction of a tree and
the legend 'JESUS NAZERENUS'
in crude lettering around the outer
edge; a fourteenth-century lead/tin
seal stamp of Nicholas de Wickes,
bent into a roll to prevent re-use;
a fragment of a German stoneware
jug with applied heraldic medallion;
pottery sherds from Italy or Spain,
extremely rare in medieval London;
a diamond knop pewter spoon,
the most common of the late-
medieval forms.
Andy Chopping, MoLAS

Perhaps the most impressive find was the discovery of numerous fragments from a pre-Industrial Revolution river vessel dated to Shakespeare's time, a barge or lighter of the type often shown on old riverscapes. Considering the site's proximity to the Globe theatre, this discovery was particularly interesting. Large parts of the boat were reused in the later revetments and placed in the dumps. Fragments of a second barge dating to the eighteenth century were also recovered. Other finds of note included a fine collection of decorated clay tobacco pipes and a number of stamped lead seals used to mark bales of cloth for trade.

The south-east quarter of the north bank site was excavated in 1974–6 as part of the Trig Lane site. This excavation set a standard for waterfront archaeology and the chance to expand our understanding of the data retrieved 25 years ago was eagerly anticipated. One of the research aims was to discover how well material excavated at Trig Lane and then reburied was preserved.

In the twelfth century the whole site would have been beneath the Thames. The earliest features found were massive tie-back structures, land anchors buried in the reclamation dumps, strengthening a revetment, which was not excavated as it lay outside the archaeological trench. The well-preserved timbers displayed remarkable carpentry skills in the formation of complex joints. On the eastern side the revetment turned to the north creating a narrow inlet in the riverfront apparently dividing two properties. Two more extremely solid revetments were excavated further to the south, indicating that during the thirteenth century approximately 12 to 14 metres of land was won from the river. It seems the inlet was retained and the eastern side of it was discovered, showing it to be barely two metres wide.

The river bank was transformed in the fourteenth century. A new high-status masonry river wall was constructed on the western property. It formed the western edge of the inlet, running more than twenty-two metres northwards and replacing the earlier revetments. The east side of the inlet was still timber at the time and evidence shows that the whole inlet was flooded daily by the tide. Oddly, the wall was not part of a building at this time, the reclaimed land behind it being covered in a series of external surfaces.

In the fifteenth century the riverscape changed again and an even more substantial wall was built ten metres out into the Thames on the eastern property, creating a large dock with the earlier wall. Eventually this dock silted up and the frontage moved south again, beyond the excavated area. The inlet was also filled in and became a lane leading down to the waterfront, known as Boss Alley on old maps. By this period buildings occupied both sides of the lane and the upper levels of archaeology revealed post-Great Fire industrial activity.

An exceptional collection of finds from the north bank included fine medieval metalwork, a quantity of leather shoes, scabbards, belts, laces and other clothing accessories (some highly decorated), imported and domestic pottery and decorative tiles.

Post-excavation analysis over the coming months will considerably refine the data retrieved from the sites. It is inevitable that we will discover even more with further work on the records, finds and samples.

Planning and Funding
**Savas Sivetidis, Judith Mayhew
and Malcolm Reading**

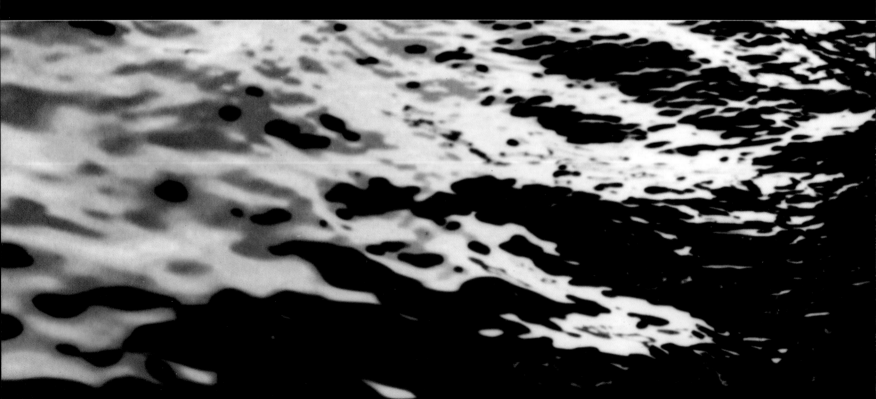

Savas Sivetidis
Director, Cross River Partnership
Formerly Head of Planning and Regeneration,
Southwark Council

Despite its central London location, Southwark has a high proportion of unemployed and socially-excluded people, and thus a key aspect of Southwark's regeneration strategy is to expand central London functions, wealth and employment opportunities south of the river. As Head of Planning and Regeneration for Southwark until January 2001, that was one of my key areas of responsibility and I have been involved with the promotion, development and implementation of the Millennium Bridge project since 1994 when the Tate Gallery decided to re-locate its modern collection in Bankside.

Southwark saw the Millennium Bridge as the most important component of its Bankside regeneration strategy. Symbolically, it would bridge the historic gap between the prosperous north and the poor south banks of the Thames. In practical terms, the bridge would make it easier for visitors to cross over from one of the main tourist attractions, St Paul's Cathedral, to Tate Modern and the Globe Theatre. With more visitors staying for longer on the south bank, Southwark anticipated increased spending and considerable secondary job creation in the hospitality industry. Thus the Millennium Bridge has always had strong support from all members and officers of Southwark Council, of whatever political affiliation, and from the three leaders of the Council in the period of the project: Jeremy Fraser, Niall Duffy and Stephanie Elsy. Southwark Council's Chief Executive, Bob Coomber has likewise been committed to the idea from the start, and has played a significant role at critical stages. My task has been to coordinate our involvement internally and to build other 'bridges' with outside organisations and individuals to facilitate the successful completion of the project, a job that has given me enormous personal satisfaction.

In the early days of the project, certainly, there were obstacles. Funding had yet to be found and the procedural routes to achieve the necessary consents seemed lengthy and complicated. Yet there was constant encouragement from David Bell and his team at the *Financial Times*. David was calm, enthusiastic, knew how to find his way around the obstacles and how to promote the Bridge effectively. I joined his steering group and worked with Dawn Austwick of Tate Modern on the development of the planning and technical brief we needed to run the design competition.

At that time I was also working on the building of another 'bridge' – the Cross River Partnership – with colleagues from the City of Westminster, the Corporation of London and Lambeth Council. In 1995 Southwark, through the Cross River Partnership which had adopted the bridge as a flagship project, obtained a government grant of £1m to develop the technical aspects of the project and obtain the necessary statutory consents. This was our first serious money and helped to accelerate progress. This strengthening of links between Southwark and the City Corporation also had a major boost early in 1996 with the first ever joint meeting of their respective planning committees to consider the planning and technical brief.

Then an event in December 1996 led to a major change in the attitude of the project's sceptics, and reduced the size of the obstacles: the announcement of the winners of the design competition. We had an exciting and beautiful design, and a design team with a worldwide reputation. The bridge had turned from an idea into something real. Now we had to organise its delivery.

With help from Dawn Austwick, I appointed Malcolm Reading as Project Director, the first person to work full-time for the client side of the project. Southwark's lawyers and planners started work on the consent procedures and the contractual aspects of the scheme. David Bell built the Millennium Trust to strengthen the promotion of the project and to assist in finding the considerable amount of money still required. The Trust, with assistance from Malcolm, appointed Carole Patey as a full-time fund-raiser and Trust Secretary. It is remarkable that for this £18m project of high political, organisational and technical complexity, the client team remained small: Malcolm, Carole and, later, Craig Bradley (Project Manager), were the client's only full-time officers. This is explained partly by the talents and competence of the three individuals and partly by the enthusiasm and commitment of many other key people.

So the bridge was built almost on time and almost on budget – a major achievement for a project of such complexity. And then it moved. The client team came back together with increased support from Bob Coomber, Southwark's Chief Executive, Andrew Colvin and Loretta Jennings of the Corporation of London, who spent a lot of their time negotiating the interface between the technical, legal and financial aspects of the remedial works. It was not now the heroic period of building a bridge out of nothing. However, the disappointment of the technical problems was only matched by our determination to complete what we had started. I moved to a new office with Craig and spent almost half my time working on the way forward for the remedial works. A difficult job but it was done.

People in Southwark Council are rightly proud of their involvement in the building of this magnificent bridge. For me it is a shining symbol of creating bridges between organisations and individuals. Southwark, through being involved in the project, made friends and built partnerships with other local authorities, the private sector and central government. The work on the bridge raised the borough's profile, and proved the Council's ability to work with organisations beyond the Southwark boundary. The bridge has closed the gap between the prosperous north and the poor south sides of the river, has drawn the south into central London, and transformed central London by recognising the river at its heart.

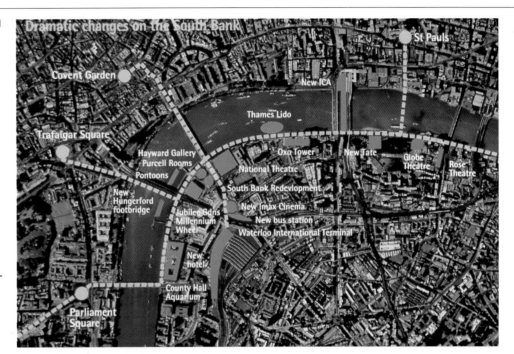

The Millennium Bridge will play a major role in the environmental and economic regeneration of London's South Bank and the Bankside area of Southwark.
Foster and Partners

Bankside Economic Study 1995
Bankside Bridge: Would It Work? 1995
MORI Poll 1996

Bankside Economic Study
Extract from a document prepared by Martin Caldwell Associates for Southwark Council and the Government Office for London (GOL), January 1995.

Bankside visitors must be understood in the wider context of visitors to London. Unfortunately, the only data which came to light on awareness of Southwark in the general London visitor was negative. In spite of the market's positive perception of the Borough attractions, Southwark has a negative image. Broadly, the breakdown of visitors to London is as follows:

- London attracts 17 million visitors per annum. Of these, 1.9 million are from the USA, 845000 are from Germany, 800000 are from Eire and 782000 from France. Numbers drop substantially after this point with Japan, as next in line supplying approximately 400000.
- Business visitors make up 33% of visitors to London but they generate half of the city's tourism revenue.
- Both business and leisure visitors show a marked preference for historic sites (75%).
- Other visitor interests were dictated by length of stay in the capital, although almost 100% sought arts and culture activities. The average stay of visitors to London was five nights.
- Other visitor interests relevant to Bankside were as follows: 50% sought art galleries, 50% sought an evening in a pub and 20% sought plays.

The London visitor profile must also be considered in the light of what we know of potential visitors to the TGMA [Tate Modern]. As the TGMA is forecast to attract as many as 2 million visitors per annum, these visitors will dominate the patterns of visitor behaviour in Bankside.

Potential visitors to the Tate are likely to share the following characteristics:

- Tate visitors break down evenly into all four age groups, not counting children.
- ABC 1 visitors predominate, although the most marked characteristics of visitors is their tertiary level of education.
- The number of repeat visits received by the Tate is higher than the average for museums and art galleries (53% once a year, 27% two to three times a year, 20% more than three times a year), a fact linked directly to the free-admission policy and the temporary exhibition programme.
- The average stay is two hours, seasonal peaks and troughs have been flattened due to good management of temporary exhibitions.
- Tate visitors expressed the following interests: 55% want to visit another gallery or museum, 23% want a sit-down meal, 22% want light refreshment, 18% want a snack, 18% want a pub, 14% want to go to the theatre or a concert, 14% would like to purchase gifts.

Visitors will come to TGMA regardless of its location. Strategies for visitor-led regeneration therefore focus on lengthening the stay of Tate visitors in the Bankside area. However, visitors to London are increasing at a rate of approximately 4% per annum. Bankside must play a part in attracting them rather than relying simply on the 'Tate-effect'. The market strategies we can draw from these broad London statistics are that the Bankside destination should be marketed in the USA, Germany, Eire and France, that the historic nature of the Bankside site must be preserved and marketed and that the interpretation provision must be of a sufficiently high calibre to attract both the tertiary education Tate visitor and the higher-spending London business visitor.

Furthermore, a recent visitor survey indicates that visitor perception of Southwark remains negative despite the popularity of Southwark attractions, and therefore projects that will improve the environment and the infrastructure will be important in boosting Bankside's image as a destination. Phase 2 & 3 linkages with large nearby attractions will be essential.

The above statistics are derived from the Tate, Southwark Council, the MVA tourism and Cultural Survey of Southwark and the London Tourist Board.

Bankside Bridge: Would It Work?
Extract from the feasibility study prepared for the Trustees of the Tate Gallery by Space Syntax, Bartlett School of Graduate Studies, University College London, 6 March 1995.

SUMMARY OF FINDINGS

The Current Study
Space Syntax was approached by the Trustees of the Tate Gallery to make an independent study of the benefits of a new pedestrian footbridge at Bankside. It was from this study, carried out over… two months in 1994, that several crucial questions arose:

1. Is the bridge really needed?

2. Will the bridge benefit the City and Southwark?

3. Would it be well used?

4. Would it make a significant difference to the areas north and south of the river linked by the bridge?

5. How best might planned developments in these areas be designed to take advantage of this potential?

6. Would it improve the future development potential of these areas?

7. Would there be a conflict between the advantages of improved accessibility and disadvantages of restricting views?

The answers to these questions are presented here in the summary of findings…

1. Is the bridge really needed?
Yes. The existing structure of space and accessibility of the south Bankside area will not be sufficient to support the major new facilities proposed for the area. It will be important to improve local area-to-area connections, including those to the north, in order to ensure a good mix of different kinds of movement and space use in the area. If these improvements are not made, there is a strong danger of another 'South Bank'; that is, a failure to link major facilities to a workable urban structure, but with an even lower density of use than the South Bank. This will almost certainly lead to demands for large-scale rectification in a few years time, which has proven to be extremely expensive in the long term, just as has happened in the Waterloo–South Bank area.

2. Will the bridge benefit the City and Southwark?
Yes. There will be immediate benefits to both sides of the river from the bridge. To the Bankside, gains will include: direct access to a major centre of working population and its facilities; a large population from which to attract support for its developments; and a potential route from the south of the Tate Modern through the riverside directly to the City of London. There will also be benefits to the City of London: direct access to a culturally significant multi-use development; increased potential of walking to work for at least some people, with increased attractions for living near the City; improved tourism potential for the City through new pedestrian linkages; and adding a liveliness to a part of the City much in need of animation.

3. Would the bridge be well used?
Yes. The evidence is that the bridge could be well used, in three ways:
a) by attracting tourists and visitors to a viewing platform half-way across the bridge, as currently happens on Westminster and Tower Bridges;
b) by attracting a proportion of these across the river to the attractors on the south Bankside, as happens now on Tower and London Bridges;
c) by creating a strong area-to-area link for background movement across the bridge.

4. Would it make a significant difference to the areas north and south of the river linked by the bridge?
Yes. The proposed footbridge would be the best possible improvement to the accessibility of the south Bankside area. Computer analysis shows this to be a powerful area-to-area link, which would radically improve access to the new cultural facilities in the area but also improve access to the area as a whole, with improvement both to background movement levels and the structuring of movement patterns in the area.

5. How best might planned developments in these areas be designed to take advantage of this potential?
The Tate building should be designed to take advantage of the new bridge by extending the line of movement created by the bridge into the building envelope, and link through the building southwards making links to the axis of Great Guildford Street. This will create a new relation between the area to the south of the power station, and points beyond, to the riverside and the north bank. This will bring a significant improvement in the local area structure, and open the way for others.

6. Would it improve the future development potential of these areas?
Yes. By making a crucial area-to-area connection, the new bridge will improve the development potential not only of the south Bankside but also the Peter's Hill area to the north of the bridge. All in all, the most important benefit from a new bridge would be the long-term improvement in area structure and therefore development potential in the areas, both north and south of the new bridge. The resulting gain in value will, in the long term, easily outweigh the cost of the bridge.

7. Would there be a conflict between the advantages of improved accessibility and disadvantages of restricting views?
No. For optimizing movement, there should be a single line of movement down Peter's Hill and across the river. But this does not mean that the bridge would need to block the view up Peter's Hill to the Cathedral from the river. The bridge does not need to be centrally aligned with Peter's Hill, provided there is a line of sight down the steps and onto the bridge. This can be achieved by offsetting the bridge to the eastern side of Peter's Hill so that there is a metre to a metre-and-a-half of overlap with Peter's Hill. This would mean that travelling in either direction on the river, the view up the steps would be virtually unchanged. With the new views from bridge, there would be a considerable net gain in visibility for St Paul's.

MORI Poll
City Attitudes to the Proposed New Pedestrian Bridge.
Summary (by Lyn Roseman and Roger Stubbs) of the research study conducted by MORI for the Financial Times, November 1996, © MORI/9768

There is clear support among people who work/have regular dealings in the City for the building of the proposed new bridge from the City to Bankside. Two-thirds (66%) support the proposal. Virtually no one opposes the plans.

One in three people (34%) who work in the City currently go across the River Thames to Bankside at least once a week; half (51%) go at least once a month. Those who go to Bankside at least once a week work in locations all over the City, with those in EC3 being a little less likely to go than others.

Knowledge about the proposed bridge is limited at this stage. Just under half have heard about the plans to develop a new bridge.

Half say they would be likely to use the new bridge to get to Bankside.

Three in ten (28%) believe they would use the new bridge at least once a week to go across to Bankside and over half think they would do so at least once a month – again drawing people from all over the City.

Judith Mayhew
**Chairman, Policy and Resources Committee,
Corporation of the City of London,
and Chairman, Cross River Partnership**

The Corporation of London is involved extensively in economic development and regeneration in the areas around the City. We are one of the richest districts in the world and yet, at the same time, are surrounded by some of the poorest areas of London and the UK. We regard that as economically, socially and politically unacceptable. Therefore, the Corporation of London is active in many partnerships with boroughs and communities which surround us. One such collaboration with the adjacent councils is the Cross River Partnership, which links the Cities of Westminster and London on the north bank to the boroughs of Southwark and Lambeth on the south. The focal point of the Partnership is the River Thames, which in the past has been seen as a geographical barrier. The Millennium Bridge is a symbolic linking of the south with the north; of the diocese of Southwark with the diocese of London; of Southwark Cathedral with St Paul's Cathedral; of the South Bank Centre, the Globe Theatre and the new Tate Modern with the Barbican Arts Centre, the Art Gallery in the Guildhall and the Museum of London. There will be a continuous north–south cultural trail linking Bankside with Cityside. The Corporation decided to make the Millennium Bridge one of its major millennium projects. This new bridge provided an excellent opportunity for the Corporation to mark its support for the regeneration of inner London and its neighbouring boroughs and to turn the Thames from a barrier at the edge of the City to a link between the City and its neighbours. The bridge also won support because the public would be able to stand on it and look at St Paul's for as long as they wished, rather than just get that momentary, 90-second passing glimpse from a boat travelling on the Thames. This quickly allayed the concerns of some that the bridge might partially obscure views of St Paul's from the river.

**A new view of St Paul's Cathedral
– from the deck of the Millennium
Bridge. St Paul's is now just a
seven-minute walk from Bankside.**
Ove Arup and Partners/Grant Smith

52

There were several processes to go through in the City of London to obtain planning permission. The proposal had to be considered by a heritage consultative group in the City. It then had to go to the full Planning Committee, where Norman Foster made a brilliant presentation, as he always does. Because it was such an important scheme, it was then reported to the full Court of Common Council, where it passed with a comfortable majority. At this point, no funding had yet been raised from the Corporation, which had been acting purely as a planning authority. The next stage was to discuss how the Corporation could contribute to the costs of the bridge. The Corporation wanted to use its charitable arm, Bridge House Estates Trust, to make a significant contribution towards what it saw as a major feature which would play a key role in the regeneration of Southwark.

Bridge House Estates Trust is responsible for the maintenance and upkeep of the four city bridges, with a history stretching back over 800 years. More importantly, in the mid '90s, a Statutory Instrument (cy pres scheme) was passed by Parliament allowing surplus funds to be given as grants to charitable organisations across London. So with the Trust we had both a funding source and a natural link to the world of bridges. When the application for funds came to the Bridge House Estates, it had to be reported to the full Court of Common Council again, where it was debated, and finally passed.

The north landing of the bridge falls into the Queenhithe Ward, which is, in fact, the Ward I represent. Indeed, I live on Queenhithe Dock, named after Eleanor of Aquitaine. When she married Henry II, she brought with her as a dowry half of France, the whole of Brittany and western France down to the Pyrenees, including Bordeaux and Gascony. The Bordeaux wine trade landed all its wine here at Queenhithe Dock, first built by Alfred the Great. When Alfred became king of Wessex, London was actually outside the Roman walls. In building this dock, he brought it back within Roman walls, the only Saxon dock in the UK that is Grade 1*. Nearby is St James Garlickhythe, the first church that Sir Christopher Wren built after the Great Fire, known as Wren's Lantern because it is beautifully light inside. So here we have Wren's Lantern side by side with the contemporary 'Blade of Light', a fitting coincidence.

One of the marvellous developments in recent years has been the way in which the South Bank, particularly Southwark Council, has developed its riverside walkway. We are also about to finish the last part of our walkway on the north, which begins at the Embankment and continues down past Blackfriars and the Millennium Bridge. As the Millennium Bridge was being built, even its greatest sceptics became proud of it and began to claim it as their own. Not only does this bridge link the two artistic districts north and south of the Thames, it also links the north and south riverside walkways in the shadows of the Tate Modern and St Paul's: a walkway of the world.

Malcolm Reading
Millennium Bridge Project Director

In January 1997, having been appointed by Southwark Council, I first met with David Bell at the *Financial Times*. The Millennium Bridge didn't have a client at this point. It had been launched publicly with the RIBA competition, at the end of which David had a great design by a 'dream team' architect, engineer and sculptor, but there was no clear plan of how to raise the funds to build it, or how to get it built. He was looking for someone to take this unusual project through the funding and planning phase, which is where I got involved. I was particularly interested in two aspects of it. First, the urban regeneration that would result from the construction of the bridge: the idea that such a tiny utilitarian object could change whole swathes of London. Secondly, this bridge was a unique fusion of architecture, engineering and sculpture. The tradition is that engineers build bridges, and occasionally an architect will be invited along to tidy up the details. And I've never heard of an artist being involved in a bridge.

After I joined the team, our challenges were twofold. First of all, I had a client that didn't exist – literally. We had various people at Southwark and various people at the *FT* and others, but we had no organizational glue to hold it together. Secondly, we needed to approach the Millennium Commission for funding and had to prove that we were an organization capable of executing this project. So my job was to co-ordinate the inception of the Millennium Bridge Trust, to devise a management system for the project partners and to meet difficult deadlines for the Millennium Commission. The application to the Millennium Commission resembled most applications to any funding body: a detailed technical review of the project, demographics, budgets, surveys, etc. Crucially, it was also a detailed review of the client, asking, for example: Is the client able to complete this project? Do they know what they're doing? Do they have their systems in place? So the technical review for our grant application was quite complex. The rest of our report to the Commission was about the management structure and how the client would finally deliver the project. Southwark Council became the statutory body responsible for building the bridge. The Millennium Bridge Trust became the project partner chiefly responsible for raising sufficient funds to match the £7.2 million grant from the

Millennium Commission. Southwark elected the contracts and took the risk and the liability for the project. Savas Sivetidis of Southwark Council was the key person there, responsible for the origins of this project along with the Trust Chairman, David Bell. They were like the yin and yang of the bridge. Savas has worked for the local authority for a long time, but his focus is on Cross River Partnership issues and Single Regeneration Budget proposals – all of the things that Southwark is trying to do so well: raising the profile of the borough and using government money effectively.

During the summer of 1997, even before we got the money from the Millennium Commission, we looked carefully at how to seek approval to construct the bridge once funding was in place. In this country, there are three ways you can build a bridge, or a railway, or a port, or a major piece of infrastructure. You can get an Act of Parliament or a Transport and Works Order – both complicated procedural routes and hugely expensive. Instead of either of these, we obtained a licence from the Port of London Authority to put a structure across the river and then sought planning permission from each of the two local authorities concerned.

We then had dozens of separate permissions to seek. With an Act of Parliament or Transport and Works Order, all the necessary permissions would have been granted with the final permission – we wouldn't have had to seek each one individually. The route we chose announced to people that this was going to be a consultative process and that we sought consensus. It transformed the way the bridge was presented to the public.

Steel sculptures by Sir Anthony Caro (the HSBC Gates) mark the bridge approach from Queen Victoria Street
Foster and Partners/Nigel Young

The Millennium Bridge has always been identified as an act of public good, which meant that we could attack some of the planning issues head on, while moving some of the technical issues to one side. It became a forum that allowed us to talk with people rather than argue. Andy Bow of Foster and Partners, in particular, made dozens of public presentations to groups of all sizes: residents' groups, community groups, disabled groups, cyclists, environmental agencies. We collected a body of opinion-formers in the City and between November 1997 and June 1998 we gave two or three presentations per week.

It wasn't until we had a contractor on board with a fixed price and a fixed programme that we felt the bridge would surely be built. I suppose the big moment was getting planning permission from the Corporation of London; from that moment on it was possible to build the bridge. That was the official green light. Southwark gave permission quickly, but the City approval was key to unlocking additional funding, and, I think, the ability to complete the project.

HSBC announced their funding two weeks before the final planning meeting in the City. This was a carefully engineered date, of course. HSBC were prepared to show their confidence in the project by announcing their financial support even before planning permission had been granted. It was a big risk for them. So the HSBC contribution tipped the balance in many ways. Immediately after that, of course, the Corporation of London made a generous contribution to the major funding package.

A third challenge was to find a way to manage this highly charged design team of architects, engineers and an artist, which consisted of people with vast experience on construction projects around the world and who are used to leading. Sir Anthony Caro came into this knowing little about the construction and planning process. It has been a shock to him, I think, how many approvals are required to get a structure built. It was my job to make sure that this development process was balanced for all the parties involved and that the parties on the project communicated properly: the design team and the contractor, the Trust and Southwark Council. All of us, I think, felt that the bridge represented a part of London's history. The bigger picture held people together, even through times when we were not sure the bridge would be built.

An enormous amount of work went into the design and planning stage long before any construction began. Almost every department of Ove Arup has contributed somehow. Their detailed analysis has been staggering and completely unseen by anybody outside of the project. The soil investigations, structural analysis, the study of the water and air flow around the bridge: all of these tasks and countless more were led by people who have immense skill and knowledge. But all of this is hidden, and in the end, of course, all you see is a slim, elegant structure across the river.

I was standing with my son, Guy, and David Bell when the bridge opened on June 10th, witnessing the opening of the sponsored walk. We saw, then felt the whole bridge begin to sway and for a moment I thought we had imagined it all. Sadly, not. After three days we had to close the bridge to pedestrians to allow Arup access for testing and analysis. Despite having a plan to manage controlled access we could not in the end get agreement to reopen the bridge and it remained closed through the year.

Finding a solution to the problem was an enormous task for Arup but they did this with terrific professionalism and energy. The ensuing design neatly conceals a series of dampers to limit the bridge moving when people are on the deck and the overall appearance is little changed.

Opening day, 10 June 2000
Foster and Partners/Nigel Young

Millennium Commission Funding Application 1996

*Millennium Commission
Funding Application
Extracts from the application to the
Millennium Commission for funding
the Millennium Bridge, prepared by
Savas Sivetidis and Malcolm Reading
and submitted by Southwark
Council on behalf of the Cross River
Partnership, November 1996.*

SUMMARY

Project Timetable

As part of the brief preparation
process, the first-ever joint meeting
of the planning committees of the
London Borough of Southwark
and the Corporation of the City
of London was held in January 1996.
This meeting approved the technical
brief in advance of the competition
being further developed.

The [bridge design] competition
was launched on 15 July 1996 and
a record number of entries was
received, over 220, at the first
stage. A shortlist of six entrants
was announced in October and
the competition winner will be
announced on 10 December 1996.

An exhibition of shortlisted
proposals, sponsored by Sainsbury's,
will open in January 1997.

After the selection of the winning
design team, a project management
team will be appointed and designs
developed for required applications
and orders from May 1997.

The final brief for the detailed design
will be issued to the design team
in October 1998, and construction
will begin on site in June 1999.

The new Millennium Bridge will
be completed for a grand opening
in August 2000.

A. PROJECT OVERVIEW

Following many years of neglect
and economic decay, the Thames-
side area known as Bankside, in
Southwark, has been recognized
as possessing great potential
for significant economic and
environmental regeneration in the
heart of the nation's capital. The past
two years have witnessed a coming
together of different agencies and
organizations with responsibility
and commitment to central London
intent on making the most of the
unique opportunities which
Bankside and adjoining areas
along the Thames offer.

Two events in particular have had a
substantial effect on the Bankside
area and the possibility of a new
bridge across the River Thames.

Firstly: the decision by the Trustees
of the Tate Gallery to create the
new Tate Modern in the disused
Bankside Power Station. This
landmark proposal was endorsed
in October 1995 by the Millennium
Commission's decision to award
the project £50 million. A four-year
development programme for the
new gallery is now underway.

Secondly: the formation of the Cross
River Partnership of central London
authorities and agencies created a
new climate of joint action to address
problems of imbalance in the heart
of the capital and represented a
counterpoint to the decision by the
Tate's Trustees.

The Cross River Partnership
was formed in 1994 and was
publicly launched in May 1995.
It comprises local authorities
(including the Corporation of
London, the London Boroughs
of Lambeth and Southwark
and the City of Westminster),
major local employers (under
the auspices of the South
Bank Employers' Group) and
organizations with strategic
interests in the area (including
the London Tourist Board,
London Transport, the Port of
London Authority, Railtrack plc
and CENTEC).

Central to the Partnership's
agenda was the need to improve
links across the River Thames,
an action incorporated into a
planning strategy endorsed by
the Partnership in January 1995,
which included the following
key elements:

- the need for distribution and
  dispersal of activities on the north
  side to relieve 'overheating' and
  to assist the regeneration of the
  south side;
- the emergence of the river as
  the integrator of the area,
  providing for the enhancement
  and sustainability of riverside
  development and activity;
- improvement to north/south
  linkages across the river to
  establish connections between
  the West End and City with South
  Bank and Bankside.

The Partnership was successful
in securing funding under the
first round of the Government's
Single Regeneration Budget
(SRB) Challenge Fund for
projects to upgrade Hungerford
Footbridge and to investigate
ways to improve transport links
across the river. In December 1995,
the Partnership announced a £20-
million programme of investment
centred on Bankside, arising from
a successful bid under Round Two
of the SRB Challenge Fund.

The objectives of the Cross
River Partnership, and of other
organizations involved in
regenerating the Bankside area,
have been informed by a number
of studies which have highlighted
the principle of improving cross-river
linkages, including the following:

- The Bankside Economic Study
  (Martin Caldwell Associates,
  for the Government Office for
  London and the London Borough
  of Southwark, January 1995)
  identified the close proximity
  of St Paul's and the Tate Modern
  as a unique combination with the
  potential to extend the average
  visitor stay in each area from two
  hours to over half a day;
- The Bankside Transport Study
  (Urban Initiatives, for the
  Government Office for London and
  Southwark Council, October 1995)
  highlighted various needs for
  Bankside, including the promotion
  of linkages by improving
  pedestrian routes and facilities,
  and the stimulation of waterfront
  and cross-river activities;
- The Thames Strategy (Ove Arup
  and Partners, for the Government
  Office for London, April 1995)
  recommended reinforcement of
  focal points of activity, raising
  standards of urban design and
  architecture and encouraging
  cross river pedestrian access.
  The area covering St Paul's
  and Bankside was noted as a
  key potential focal point on the
  Thames. The Strategy highlighted
  the opportunity to provide a new
  pedestrian bridge and, thus,
  realize the focal point;
- 'Bankside Bridge: Would It Work?'
  (Space Syntax, The Bartlett
  School of Graduate Studies,
  University College, London, for
  the Trustees of the Tate Gallery,
  March 1995) assessed the
  pedestrian use of river crossings
  in Central London and concluded
  that a pedestrian bridge would be
  a well-used link for tourists and
  visitors to the area.

These studies, reports and strategies have all identified that there have, over the years, been numerous schemes and proposals for a bridge across the Thames at Bankside. In re-affirming [this] concept, the authors have recognized the high level of popular and professional interest in the idea of a new bridge. Every visitor to the Tate Gallery of Modern Art will see the imposing edifice of St Paul's a few hundred metres across the river. Much of the enthusiasm to build a bridge has arisen from the natural desire to reinforce the visual link between the two buildings and to compound the growth in visitor numbers which will be generated by the new gallery in particular. …

A.2 The National, Regional and Local Significance of the Project

… In the past three years there has been substantial development interest in the South Bank for a wide range of new visitor attractions and improvements to existing facilities. The following examples indicate the potential level of investment and visitor activity in the South Bank area and the context within which a new bridge will be set.

Project	Cost (£m)	Estimated Annual Visitors
Tate Gallery of Modern Art	130	2 million
Shakespeare's Globe Theatre and Exhibition	29	750 000
South Bank Centre Redevelopment	127	6.5 million
Imax Cinema complex	20	750 000
Potter's Field Theatre and Open Space	20	480 000
Oxo Tower Wharf	20	300 000
TOTAL	346	10,780,000

A.4 Public Support for the Project

… The Bankside Residents' Forum, the local residents group in the Bankside area of north Southwark, has the stated policy of encouraging the development of pedestrianised walkways. It is particularly keen that Bankside 'should… be made more user-friendly' (Bankside Residents' Forum Principles, January 1996). The Bankside Residents' Forum and South Bank Forum have received presentations on the project along with regular updates on progress. The Financial Times held a well attended consultation event for local residents, in July 1996, which revealed support for the bridge as a desirable and useful feature for the local residential community. Residents have expressed a requirement for a design which is of the highest quality [with] maximum accessibility for unrestricted and safe use by local people. …

A.5 Creative Involvement in Design and Construction

… The Financial Times announced in its pages the two-stage competition on 15 July 1996. A copy of the competition conditions is supplied as a free-standing annex to this application. The objective of the competition was as follows ('Millennium Bridge Competition', Financial Times, July 1996): 'to promote a new pedestrian bridge across the River Thames between Bankside and St Paul's. The bridge should be of exceptional quality both in terms of its design and engineering solution, offering

The first stage of the competition was open to registered architects, engineers and artists throughout the world and particularly aimed at encouraging the UK's best designers. The first stage required outline designs from the entrants. Submissions closed on 25 September 1996 and over 220 entries had been received. A panel of judges convened in the week following submission and began the task of selecting six entrants who would be invited to work up their schemes for the second-stage submission. The list of the judging panel is contained in Annex 9. The shortlist of six designs was announced on 7 October 1996, and finalists have been given an honorarium and five weeks in which to refine their design proposals for the second-stage submission deadline on 15 November 1996. The names of the six design teams and photographs of the shortlisted designs are contained in Annex 10. Following the second stage assessment in the week beginning 18 November 1996, the winner will be announced on Tuesday 10 December.

The panel of judges has been supported in its deliberations by a technical panel which comprises experienced experts in the fields of architecture, engineering and planning. The technical panel has advised that all six shortlisted schemes appear structurally feasible and could be built within a £10 million budget. The list of the technical panel is contained in Annex 9 to this application.

A.6 Number of Users

The bridge will be located in the heart of an international visitor destination. The area will be supported by transport links which include the international rail terminus at Waterloo, four main-line railway stations (London Bridge, Cannon Street, Waterloo and Blackfriars), five existing underground stations, plus two new stations in Blackfriars Road and London Bridge when the Jubilee Line extension opens in 1998, and a range of bus services.

Estimates of bridge use can be gauged by a combination of attendance projections at the growing range of local visitor attractions and the further development of a sophisticated public transport infrastructure bringing people into the area. Research by Ove Arup & Partners late in 1995 stated that 2 350 visitors to the new Tate Gallery of Modern Art will use the bridge every day during the week, rising to 3 800 visitors per day over the weekend. These estimates may be considered to be extremely conservative and do not take into account the local residential and business pedestrians who will be using the bridge throughout a twenty-four-hour period. There are 6 271 residents in the Cathedral Ward of Southwark, covering the local area within immediate influence of the bridge. There is a total of 32 500 employees within Cathedral Ward and a total of 229 300 employees coming into the City each working day. Taking these sources in combination, the expected number of users of the bridge has been tentatively calculated in the vicinity of 11 000 people each day, or over four million people each year...

B. PROJECT IMPLEMENTATION

...B.4 Development Timetable

The proposed development timetable for the Millennium Bridge will comprise five stages:

Stage	Activities	Schedule
Stage 1	*Design Competition and Selection* Design competition conducted Winning design team selected	to December 1996
Stage 2	*Preliminary Design and Funding* Exhibition of shortlisted schemes Public Consultation Appoint Design and Project Management Team Develop design concepts and secure funding Programme of statutory consultation	January 1997 to April 1997
Stage 3	*Consents and Approvals* Development of detailed design for applications Applications made for formal approval, consents and orders Consents achieved and orders obtained	May 1997 to December 1998
Stage 4	*Detailed Design and Tendering* Final Brief issued to design team Detailed scheme design Tender documents and specifications prepared Tender process and appointment of contractor	October 1998 to May 1999
Stage 5	*Construction* Site preparation and enabling works Construction of piers and landings Fabrication and on site erection Project completion and Opening	June 1999 to August 2000

ANNEX 1
Summary of Background Information

Section One –
Documentation produced by and on behalf of the Cross River Partnership

1.1
'Cross River Partnership Bid 1994' – submission document for funding from the Single Regeneration Budget Round One. Submitted September 1994, produced by The Cross River Partnership, co-ordinated by Southwark Council.

1.2
'Bankside Bridge: Would It Work?' – feasibility study. Circulated [March 1995], Produced by University College London for the Trustees of the Tate [Gallery].

1.3
'Cross River Partnership Framework Document' – Establishes the co-ordinated strategic framework for the Partnership. Circulated January 1995, produced by Urban Initiatives on behalf of the Cross River Partnership.

1.4
'Bankside Economic Study' – study to identify the key interventions to establish Bankside as a mixed development area. Circulated February 1995, produced by Martin Caldwell on behalf of Southwark Council and the Government Office for London (GOL).

1.5
'Competition to select an architect' – invitation and briefing document for the selection of the Tate architects. Circulated February 1995, produced by the Tate Gallery.

1.6
'Cross River Partnership SRB Delivery Plan' – document outlining the implementation of the Cross River Schemes approved in 1994. Ratified June 1995, produced by Urban Initiatives on behalf of the Cross River Partnership.

1.7
'Bankside Transport Study' – project development report to underpin the 1995 SRB bid document. Circulated September 1995, produced by Urban Initiatives on behalf of the Cross River Partnership.

1.8
'Bankside Consultation and Local Management Study' – provides the context and strategy for communicating with the various local communities. Circulated September 1995. Produced by EDAW Planning on behalf of Southwark Council.

1.9
'Cross River Partnership Bid 1995' – submission document for funding from The Single Regeneration Budget Round Two – The Challenge Fund. Submitted September 1995, produced by The Cross River Partnership, co-ordinated by Southwark Council.

1.10
'LBS/Cross River Partnership SRB Delivery Plan' – document outlining the implementation of the Cross River schemes led by Southwark Council. Approved in 1995, ratified May 1996. Produced by London Borough of Southwark on behalf of the Cross River Partnership.

1.11
'CRIPTS Paper on Travel Patterns in the Cross River Partnership Area' – report of detailed analysis of traffic/pedestrian flow in the Cross River Area. Circulated August 1996, produced by South Bank Employers on behalf of the Cross River Partnership.

1.12
'Cross River Partnership Bid 1996' – submission document for funding from the Single Regeneration Budget Round Three – The Challenge Fund. Submitted September 1996, produced by The Cross River Partnership, co-ordinated by Southwark Council.

Section Two –
Millennium Fund Documentation

2.1
'The Bankside Bridge: Celebrating The New Millennium in Central London' – application to the Millennium Commission. Submitted February 1996, prepared by Southwark Council on behalf of the Cross River Partnership.

2.2
'Hungerford Bridge – Bridging Communities' – application to the Millennium Commission. Submitted February 1996, prepared by The City of Westminster on behalf of the Cross River Partnership.

2.3
'Proposed Pedestrian Bridge Bankside – Peter's Hill' – planning and technical brief. Developed January 1996 as part of Bridge Competition Literature, July 1996. Prepared by Ove Arup and Montagu Evans, commissioned by Southwark Council.

2.4
'The Millennium Bridge Competition' – instructions to entrants. Distributed July 1996, prepared by the *Financial Times*, Royal Institute of British Architects and Southwark Council.

Section Three –
Statutory and Strategic Planning Documentation

3.1
'London Borough of Southwark – Unitary Development Plan'. Adopted July 1995.

3.2
'Strategic Guidance for Planning Authorities' – issued by Government Office for London, May 1996.

3.3
'Strategic Planning Guidance for the River Thames' – issued for consultation by Government Office for London, June 1996.

Design and Engineering
Norman Foster, Roger Ridsdill Smith
and David Newland

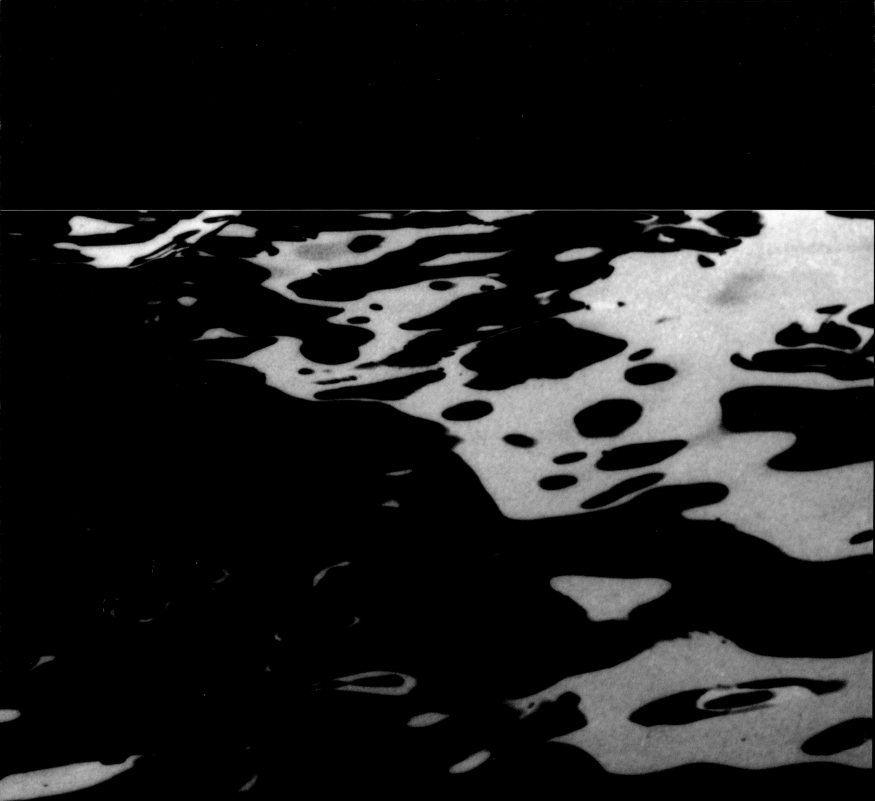

Norman Foster
Foster and Partners

Sketches by Norman Foster.
Foster and Partners

Building bridges is about bringing people together – forging links and establishing lines of communication where none previously existed. The Millennium Bridge does that in a very real physical sense – connecting one of London's richest boroughs with one of its most underprivileged. Designing the bridge has also fused the worlds of engineering, art and architecture and its construction has brought together skills and craftsmanship from all over Europe. These are the bridges that no-one can see, but they are a crucial part of the story.

The Millennium Bridge is truly a bridge for the new millennium. It will allow people to enjoy some of London's greatest landmarks from a new vantage point. It will form an essential part of London's pedestrian infrastructure: every year millions of people will use it to cross the river or simply to enjoy a totally new experience of London and the Thames. It makes some of the capital's most important landmarks, old and new, accessible to a greater public and provides the city with a remarkable new attraction. It is a place to promenade – a new public space – away from the tyranny of the city's traffic.

The bridge links the City of London and St Paul's Cathedral to the north with the Globe Theatre and Tate Modern on Bankside. It will have a social and economic impact on both sides of the river, creating new routes into Southwark – thus contributing to its regeneration – and encourage new life on the embankment alongside St Paul's. The river all too often seems like a barrier between north and south London rather than a focus of social activity at its centre. The Millennium Bridge turns the river into a connection rather than a separation. I think that is why it is so exciting and perhaps why it captures our imagination.

The Millennium Bridge springs from a creative collaboration between engineering, art and architecture. It was developed together with the engineers Arup and the distinguished sculptor Sir Anthony Caro. I am fascinated by the relationships between function, technology and aesthetics and as an architect I have always worked closely alongside sympathetic engineers. This has been a deliberate choice to break the established mode, traditional in most schools of architecture, where the student is encouraged to design in a relative vacuum, assured that the engineering professionals will later be able to translate the design into a reality. I have enjoyed a thirty-year-long relationship with Arup. We have worked together on some of our most challenging projects, including the Hongkong and Shanghai Bank and the Collserola Telecommunications Tower in Barcelona. I have known Anthony Caro for just as long, since we were neighbours in the early 1970s.

We should not forget that the sculptor, the engineer and the architect are equally concerned with gravity and grace; with volume, mass and space; and resolving and harmonising the stresses and tensions between physical forces. We often share the same tools for creating our products – Caro changed the vocabulary of modern sculpture when he started to construct his pieces with building girders in the early 1960s. It is vital to remain open to new ideas and to recognise that they can come from anywhere – and probably won't come from the hermetic world of one's own creative discipline. I am more interested in the common features of our disciplines and not in the less-important differences.

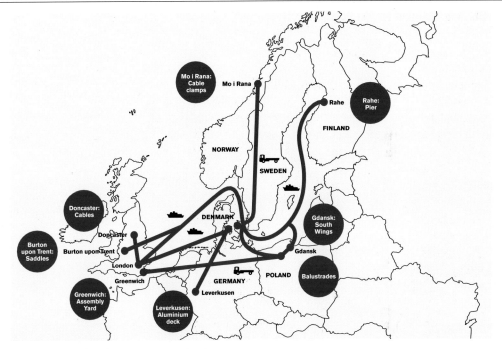

Map showing origins of components of the Bridge.

U-Pein Bridge, Amarapura,
near Mandalay. This bridge is
1208 metres long, made of teak
and was built from 1849-1851.
Michael Macintyre,
Hutchison Picture Library

When we came together as a team, the challenge was to realise something that would be elegant and minimal – an extraordinary feat of sculpture, architecture and engineering.

Infrastructure has a very powerful visual impact on the environment. Bridges often provide an opportunity to create monuments – instantly recognisable symbols of a particular place. In the Golden Gate Bridge, the Brooklyn Bridge or the Clifton Suspension Bridge the audacity of the engineering speaks for itself – it is obvious in the enormous distances spanned or in the huge height of the towers and the length of the cables that support the deck. Our task was quite different – to create a bridge so sleek that it does not fight for attention with the historical monuments for which it provides a link. There is no point in inviting people to the middle of the river to enjoy London from a unique viewpoint and then obstruct the views with the bridge's own structure.

We worked together to minimise the bridge's intervention in its setting. It is the ultimate expression of spanning the greatest distance with the minimum means. Structurally, it is a very shallow suspension bridge, spanning 320 metres. It relies on eight cables, four on either side of the 4-metre wide, lightweight deck. The cables dip just 2.3 metres over the 144-metre central span. Steel transverse arms support the deck by clamping onto the cables at 8-metre intervals. By day it appears as a thin ribbon of steel and aluminium. At night it forms a glowing blade of light.

Its construction results from an incredible multi-disciplinary collaboration across nine European countries: the cables from England, the cable clamps from Norway, the steelwork from Denmark and Poland, the handrails and decking from Germany, the pier arms and wings from Finland, the stainless steel from Sweden and the stonework from Ireland. A huge number of people worked together on the project. A conservative estimate was five thousand. I think you could easily double that! Just in terms of coordination it was an awesome task.

The bridge opened in June 2000 and 100,000 people crossed it during the first weekend – an indication of the huge public interest it had generated. As everyone knows, because of unexpected swaying the bridge had to be closed. This was a huge disappointment to everyone who worked so hard to build it and to the millions of people who looked forward to using it. In designing the bridge we employed structural technology that did not exist even ten years ago. We had to invent the means to make one of the longest pedestrian suspension bridges in the world, with a cable sag of only 2.3 metres; around six times shorter than a traditional suspension bridge.

With the closure of the bridge, we were faced with a new technological challenge – to dampen the movement without visually compromising the original concept of the bridge. The team was reassembled and the expertise of the entire international engineering profession was enlisted. Extensive research and testing have resulted in a discrete damping system that will not alter the bridge's form. We have learned valuable lessons from these investigations, which have led to the revision of the international codes for bridge building. When the bridge re-opens it will be a testament to an extraordinary technological and artistic collaboration.

**Ken Shuttleworth, Andy Bow
and Norman Foster discussing the
bridge plans at Foster and Partners'
Battersea offices**
Foster and Partners/Nigel Young

'For me, turning the concept into
reality is always a fascinating
process. The detailed analysis along
the way means that the concept is
constantly being challenged. Some
ideas get dropped but new ones
will come to the fore. It is a magical
process. I can't define exactly how
it works as each job is different.'
*Ken Shuttleworth, Partner,
Foster and Partners*

**Aerial photograph of the bridge
site, with computer simulation
of the finished bridge, submitted in
the first stage of the competition.**
Foster and Partners / Nigel Young

'In the first stage of the competition,
Norman had drawn a sketch of the
dome of St. Paul's Cathedral with
the bridge linking it to Bankside
and the new Tate Modern.
With only twenty-four hours to
go before the scheme had to be
submitted, Norman said he wanted
an aerial photograph of exactly that
image. The weather was appalling
at the time, but he insisted that
we get a helicopter in the sky to
take a photograph that expressed
the geographical significance of
this project.

So, despite the wind and heavy
rainfall, the helicopter went up
and hovered over the Thames for a
while, waiting for a shaft of light in
the distance to travel over London
from Kent. Eventually the light fell
exactly on the site for the bridge,
and the resulting photograph
became a key inspirational image
for us, which we later sent to all
the different manufacturers of
the bridge components.'
*Andy Bow, Director,
Foster and Partners*

'Weekly meetings and visits within the team were absolutely necessary. When we were completing the design development drawings for contract, Ove Arup set up a work station for me on their premises. This enabled us to work closely with the Arup engineers and with representatives from the Danish contractor Monberg & Thorsen. This kind of proximity between architects and engineers on a project is uncommon, but it worked brilliantly. The usual separation between our two disciplines became blurred at certain stages to the point where colleagues in the building profession would be surprised to hear the architects asking the engineers for opinions on aesthetic issues, traditionally the architect's domain.

Just as the architects 'untraditionally' offered aesthetic views, I 'untraditionally' travelled with the engineers to witness necessary technical testing, such as hydraulic modelling, bridge dynamics and wind-tunnel tests. I also visited most of the factory sites to inspect the fabrication of the various parts of the bridge.'
Catherine Ramsden, Project Architect, Foster and Partners

Above: Hydrological tests, modelling the effects the bridge piers have on the currents and tidal flows of the Thames; Catherine Ramsden and Roger Ridsdill Smith are pictured right.
Foster and Partners/Nigel Young/ H.R. Wallingford

Above: Panoramic photomontage.
Hayes Davidson / Nick Wood
Right: Sketches by Roger Ridsdill Smith.
Ove Arup and Partners

'In the early stages of detailed design, it was decided to tender the construction work in two stages, which meant that after an initial costing we could appoint the contractors early and get their input into the design specification while we were still drawing it up. Bridges require unusual construction techniques and much of the cost is hidden in the building methods. The contractors Monberg & Thorsen and Sir Robert McAlpine were innovative in their approach, so we were fortunate to involve them at such an early stage.'

Monberg & Thorsen were responsible for every bit of steel structure and deck while McAlpine were responsible for all the concrete and foundation work. They had full power over choosing their subcontractors, and we didn't interfere with that process. They handled the tendering of the packages and appointed companies all over Europe.'
Sophie Le Bourva, Project Manager, Ove Arup and Partners

'When you look at the deck now, its design seems so obvious, it is doin what it needs to do, every detail ha been carefully integrated and there is no clutter, not one extraneous part. That did not come easily. This simplicity is the result of several generations of deck cross-sections and full-scale checks using an evolving working prototype before we honed down the final design.'
Sophie Le Bourva, Project Manage Ove Arup and Partners

The fixed end result indicated that at 3.4 m, the load would be 15.277 kN/m.

$$T = T_0 \times 1.5 \times D_0/D = 8.64 \times 1.5 \times 3/D = 38.88/D$$

$$e_s = \frac{8}{3L} D_0^2 \left(1 - \left(\frac{D}{D_0}\right)^2\right) \cdot \frac{1}{6}\left(1 - \left(\frac{B}{B_0}\right)^2\right)$$ Sub L & D_0

$$e_T = \frac{(T-T_0)L}{EA_c} = \frac{T-T_0}{9.133E3} \times 144 = (T-T_0) \times .01577$$

$$e = e_s + e_T$$

$$k = \frac{2\Delta T}{e} \quad \frac{\Delta T}{(e/2)}$$

$$L^2 = \frac{8T_0}{\omega_0} = \frac{8TD}{}$$

$$T = \frac{\omega L^2}{8D} \quad T_0 = \frac{\omega L^2}{8D}$$

$$T. \quad ie. \quad \boxed{T_{rev} = \frac{T_0 \quad \omega_2}{L^2 D}}$$

D	T	ΔT	e_T	e_s	e	k (MN/m)
3.4	11.435	2.795	.04408	-.04741	-.00333	1680
3.45	11.270	2.630	.04147	-.05375	-.01228	428

'We had to design curved bullnose elements, which cover the light tubes along the whole length of the bridge, on both sides. Initially we wanted them to be transparent. So for almost nine months we researched plastics and laminated glass to find a suitable material which could be moulded into the slim profile we required. But the Corporation of London planners became concerned about two things, mainly because the bullnoses would be illuminated. The first was graffiti, which could be sprayed on the underside of the clear covers and would be difficult to clean. The second was gnats and moths. The Corporation of London felt – quite rightly, I think – that the bridge would become the world's greatest attraction for flying insects! Not the most inviting environment for pedestrians crossing the bridge. In the end the bullnoses became stainless steel too, but designed with apertures so that the bands of light each side of the bridge are visible both on the deck and in elevation. This change was one of the key moments in the project's evolution because we learned that we couldn't be wedded to any particular idea, that if the bridge was to succeed, we had to respond to people's objections and move on. It has been and always will be a very

Next page: Photomontage,
including original 'eye of the needle'
design for the bridge's south end.
Hayes Davidson / Nick Wood

A brief history of suspension bridges

AD 0–600
Suspension bridges have existed for more than 2000 years. The earliest were made of rope. Iron chain suspension bridges are reported as existing in China in the sixth century AD.

1800–50
In Europe, France in particular, iron chain suspension bridges became fashionable in the early part of the nineteenth century. The Menai Bridge, between the coast of north Wales and the isle of Anglesey, was one such, opened in 1826. It was damaged three times (1826, 1836 and 1839) by high winds and restored and altered after the gale of January 1839. Another dramatic failure occurred in 1850 at Basse–Chaine-Angers, when the 102-metre suspension bridge collapsed over the Maine River under the weight of a column of troops. It was in response to such accidents that William John Macquorn Rankine [1820–1872], a Scottish engineer, published *A Manual of Applied Mechanics* (1858), which includes his theory on the interaction between cables and deck–stiffening girders.

1850–1900
The latter part of this century (1880–90) saw the development of the elastic theory for the behaviour of suspension bridges and the deflection theory, which is based on the differential equations of the cable and the stiffening girder.

1900–1950
The practical effect of these theories was seen in the early twentieth century, when several major suspension bridges were built with much shallower girder depths than had been common before: the Golden Gate Bridge in San Francisco (1937) had a span–girder depth ratio of 1:168 and the Tacoma Narrows Bridge in Washington, USA (1940), one of 1:350. However, only four months after completion, on 7 November 1940, the Tacoma Narrows Bridge collapsed as a result of oscillations induced by a 40-mph wind.

1950–1970
In the 1950s, the link between torsional strength and wind stability became better understood and span–girder depth ratios increased. The Mackinac Straits Bridge, Michigan, opened in 1957 with a span–girder depth ratio of 1:100. The 1960s saw a further advance as computer analysis of non-linear structures, such as suspension bridges, allowed the development of the technique known as dynamic relaxation. The opening of the Severn Bridge in 1966 introduced another novelty: the use of a profiled closed section rather than the tradition altruss girder for the deck. The span–girder depth ratio for this bridge was 1:324.

2000
The Millennium Bridge marks a dramatic advance in suspension bridge building, with a deck that rises above and falls below the cables, with a span–dip ratio of 1:63 – six times shallower than that of traditional suspension bridges. The combination of a profiled cross-section and wide band of cables provides torsional stability and allows the span–girder depth to be reduced to a ratio of 1:443.

The Clifton Suspension Bridge, 1864
This deck is suspended from three wrought iron eye-bar chains on each side. The deck girders are 0.9-metre-deep wrought iron plate girders; the deck always falls below the chains.

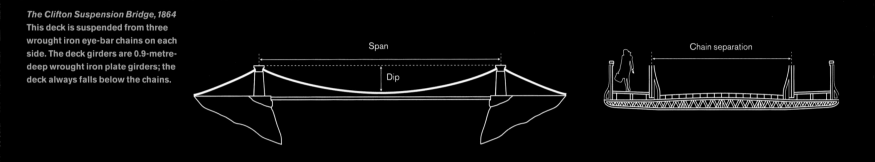

The Golden Gate Bridge, 1937
This deck is suspended on both sides from 250mm diameter steel cable of 27500 wires. The deck girder is 7.6 metres deep and is composed of four trusses (top, bottom and sides). Aerodynamic stability is from the closed deck section; the deck always falls below the cables.

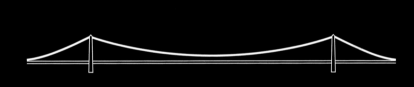
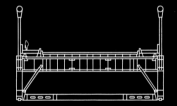

The Humber Bridge, 1981
The deck is suspended on both sides from 680-millimetre diameter steel cable made up of 15000 wires. The deck girder is a steel box 4.5 metres deep. Aerodynamic stability is from the profiled deck shape; the deck always falls below the cables.

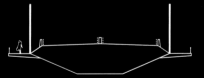

The Millennium Bridge, 2000
The deck is suspended/supported on each side by five 105-millimetre diameter steel cables. The deck girder is a 0.3-metre deep circular steel hollow section. Aerodynamic stability is improved allowing the deck to rise above and fall below the cables.

Roger Ridsdill Smith
Project Manager, Ove Arup and Partners

Competition

The Millennium Bridge started life as three lines on a page, two for the plan and the third for the elevation, the essence of a crossing, the most direct route between two banks. This appealed to the team as an elegant and logical design for a bridge sandwiched between a busy navigation channel and the height restrictions necessary to maintain a clear view to St Paul's Cathedral.

Our initial structural calculations were for a deck slung between two steel cables, and, after a brief check on a concrete alternative, this is what we settled on. The cables were to sag 2.3 metres vertically across the central 144 metre span, six times less than a traditional suspension bridge. Two piers would prop this span, and two large abutments would anchor it on the banks.

We were incredibly excited to hear that we had won the competition in December 1996. Most of 1997 was spent producing information for the planning application and preparing a dossier for Millennium Commission funding. Corinne Swain provided strategic planning consultancy and her advice on the choice of planning route saved the project an enormous amount of time.

Engineering Team

While this was going on the team who would build the bridge was being put in place. Chris Wise led the job until he left Arup in June 1999, when the foundation construction was underway and the construction drawings for the main works were almost complete. David Kaye, who had been involved with the job from a distance, took over at this point and brought with him a wealth of knowledge of civil engineering. Roger Ridsdill Smith was the Project Manager, responsible for the day-to-day running of the job, overseeing the design and administering the contract during construction. Over seventy people from ten different departments were involved within Arup, from planners and archaeologists at the beginning of the job, to geotechnicians, maritime engineers, traffic engineers, lighting engineers, mechanical and electrical engineers, and obviously many structural engineers. The structural design team was led by Sophie Le Bourva who coordinated the various engineering disciplines within Arup. The calculations for the structure were checked by an independent team within Arup as well as a separate engineering firm.

The bridge is a result of a close collaboration between the members of the design team. The engineers worked very closely with the architects on every aspect of the job. Considering the intensity of the design process it is perhaps surprising that the final form of the bridge looks very similar to the original competition idea.

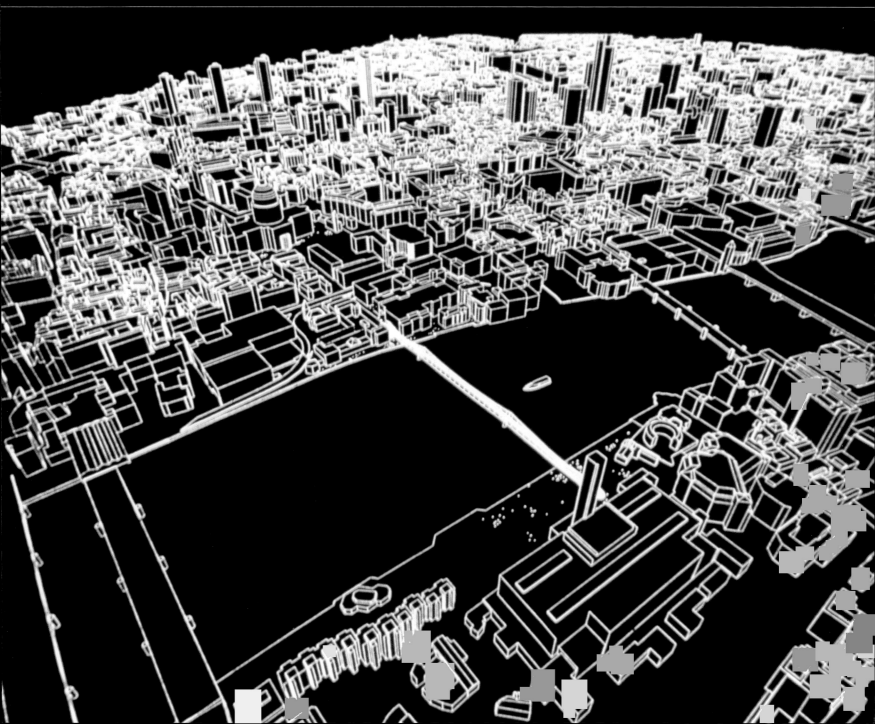

Site clearance began in December 1998, including demolition of the coal pier that used to serve Bankside Power Station.
Foster and Partners / Nigel Young

After the archaeological dig was completed in May 1999, work began on the piles for the bridge abutments, 16 on the south site and 12 on the north.
Ove Arup and Partners / Grant Smith

Superstructure

The bridge is a shallow suspension bridge, where the cables are as much as possible below the bridge deck to avoid obstructing views from the deck. Two groups of four 120 mm diameter locked coil cables span from bank to bank over two river piers. The lengths of the three spans are 81m for the north span, 144m for the main span between the piers and 108m for the south span. Fabricated steel box sections are set between the two cable groups every 8m. The 4m wide deck structure comprises two steel edge tubes which are supported by the transverse arms. The deck itself is made up of extruded aluminium box sections which span between the edge tubes on each side. The lighting and handrails are also fixed onto the edge tubes.

The groups of cables are anchored at each bank. Each abutment is founded on a 3m reinforced concrete pilecap anchored by a group of 2.1m diameter reinforced concrete piles. There are 12 piles on the north bank and 16 on the south bank where the site is constrained and the pilecap shorter in consequence. The river piers themselves comprise a steel 'V' bracket fixed to a tapering elliptical reinforced concrete body which is founded on two 6m diameter concrete caissons.

Cable design

Ribbon bridges have similar shallow profiles to the Millennium Bridge, but are typically single spans, allowing the cables to be anchored directly to substantial stiff abutments which help limit the live load deflections. The Millennium Bridge is unusual in having multiple spans. The river piers are quite slender and cannot provide stiffness comparable to that of a massive abutment. This means the spans interact, making the behaviour of the structure more complex than that of a single span bridge. For example, if only the central span were loaded, the outer spans would deflect upwards.

The shallow cable profile results in large tensions. The mass of the bridge is 2 tonnes per linear metre along the bridge axis. The resulting tension in the cables is a total of over 2000 tonnes divided between each side. This large tension provides the stiffness both vertically and laterally for the bridge.

The cables are set wide apart in plan and well beyond the width of the deck. This has two advantages over placing the cables immediately beside the bridge deck. First, the deflection of the bridge, if all of the pedestrians stand on one side only, is reduced. Secondly, the structure is more stable in high winds because the cable arrangement reduces interaction between vertical and torsional vibration of the bridge.

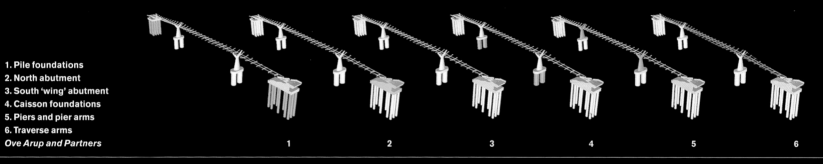

1. Pile foundations
2. North abutment
3. South 'wing' abutment
4. Caisson foundations
5. Piers and pier arms
6. Traverse arms
Ove Arup and Partners

1   2   3   4   5   6

Top: Bridge elevation.
Bottom: Bridge plan.
Ove Arup and Partners

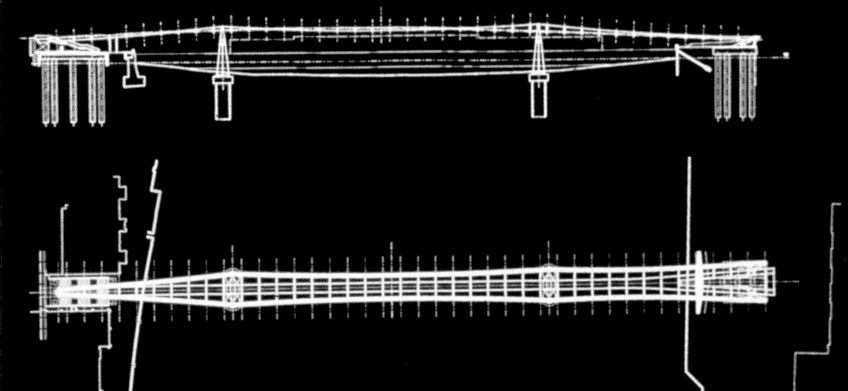

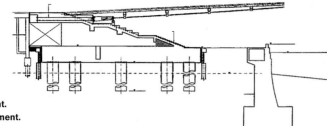 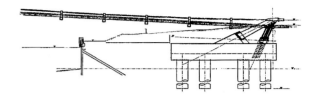

Right: Section of the north abutment.
Far right: Section of the south abutment.
Ove Arup and Partners

Above: An elaborate system of steel bars was constructed to reinforce the concrete poured into the pile cavities and abutment skeleton.
Ove Arup and Partners / Grant Smith

Construction

Construction of the bridge began in November 1998. A bank, a school, a theatre, an art gallery and many residential buildings surround the site. Access was difficult and there were strict noise controls; we were erecting a structure across a fast-flowing busy commercial river and were not allowed to interrupt the riverboat traffic.

A significant, and now almost invisible, part of the engineering of the bridge is in the ground. John Seaman led our geotechnical team and worked through a series of different schemes. The foundations of the bridge resist the horizontal force of 2000 tonnes mentioned above, equivalent to a capacity crowd at Wembley stadium pulling from both banks of the river. The bridge landing is squeezed between two buildings on the north side, both of which are sensitive to any excessive movement. In addition, the stiffness of these foundations had to be accurately calculated in order to inform the design of the bridge superstructure.

The tender to build the bridge was won by a joint venture of Sir Robert McAlpine of the UK and Monberg and Thorsen from Denmark who joined the team during 1998. The job became truly European.

The scale of the bridge resulted in some dramatic events during construction. In September 1999, more than 100 lorries from three different sites in South London carried 1000 cubic metres of concrete to the south site. A 3m thick concrete slab was cast on top of the piles. This structure now bears the full force of the cables. The north side pour followed shortly afterwards.

At the same time as we constructed the abutments on both riverbanks, we were also building the piers inside two cofferdams in the river. Cofferdams provide a dry area in which to work and are made by driving a series of interlocking steel sheets into the riverbed and draining the resulting space. Once the water was pumped out and the riverbed exposed, two caissons were dug for each pier. The caissons are six metre diameter reinforced concrete cylinders, going down 25 metres, so deep that when workers were digging at the bottom of the pit, air conditioning had to be provided by the contractor to enable them to breathe. This difficult work went smoothly and was one of the great successes of the construction. Once the concrete river piers were finished, the steel anchorages were fixed to the north and south abutments and steel brackets were attached to the top of the concrete piers in the river. Each bracket weighs about 80 tonnes and had to be lifted into place by a crane on a barge and bolted onto the pier with steel bars fixed deep into the concrete.

With the supports in place, we installed a 'highline' cable about 20 metres above the bridge. This was an innovation proposed by the contractor: a highline system is more often used to build cable cars in the mountains. We used it to draw the bridge cables from one side of the river to the other, without disturbing river traffic, or relying on the tides to allow access for floating cranes.

Right: Formwork for the reinforced concrete piers was floated up the river on barges and set in position over the caissons
Foster and Partners/Nigel Young

Above: The south abutment wings are lowered into place.
Ove Arup and Partners/Grant Smith

Above: Two enormous cofferdams were built in the river and drained of water so that two caissons for each pier could be sunk deep into the river bed.
Foster and Partners/Nigel Young

Above: On the south site, the steel reinforcement clearly outlines the position of the pilecap and plinths onto which were placed the pre-fabricated 'wings' that anchor the bridge cables.
Foster and Partners/Nigel Young

After both piers were finished in January 2000, pre-fabricated pier arms were craned into position and their bases fitted precisely onto protruding steel framework.
Ove Arup and Partners/Grant Smith

Top: Front and side elevations of the bridge piers.
Bottom: Pier plan.
Ove Arup and Partners

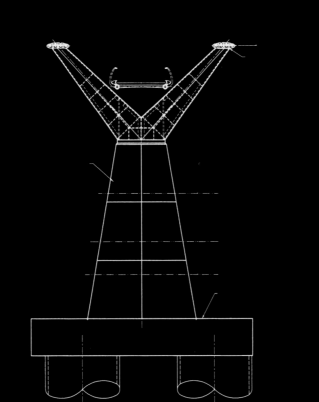

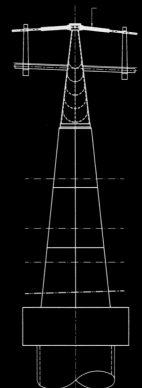

Many of the components of the
bridge were fabricated off site, such
as the cable clamps, stainless steel
handrails and locked-coil cables.
Ove Arup and Partners/Grant Smith
Foster and Partners/Catherine
Ramsden

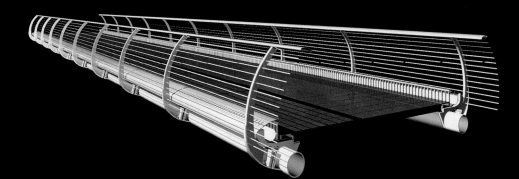

CAD drawings of deck cross-sections
showing bullnoses and light tubes.
Foster and Partners

Pulling cables across the river using the highline system
Ove Arup and Partners/Grant Smith

The cables arrived from Sheffield on reels and were pulled from the south side to the north side on the highline as if they were on a curtain rail. They were then lowered gradually onto the piers. As they were lowered, they were pulled tighter at each end so that they would land on the pier arms in the correct position. It had to be a precise operation because the cables would have been damaged if they had been pulled taut and dragged over the piers at any time.

In order to achieve the final geometry, the cables were installed with a shallower dip than required. Once all eight cables were in place, the pre-fabricated deck units arrived by barge from an assembly site in Greenwich. Placing these 16 metre long deck units on the taut cables over the river was possibly the most difficult stage of the construction. As the weight of the deck units was applied, the cables gradually sank to the correct level. This is easy to describe, but very difficult to achieve on such a complex site. The deck erection process went smoothly and the construction of the main works finished with landscaping on each abutment.

Opening day

The bridge opened to the public on 10 June 2000. Some 100,000 people crossed the bridge during the course of the day. As with all suspension bridges, the Millennium Bridge is subject to a degree of vibration. However, when this large group crossed, much greater than expected sideways vibrations occurred.

The bridge was closed on 12 June in order to investigate the reasons for the movement and to design a system to reduce it. The closure was followed by a storm of publicity. Tony Fitzpatrick, Chairman of Arup Building, announced to the world's media that Arup would find out what the problem was and resolve it.

Investigations

We now believe that the Millennium Bridge vibration was caused by the natural swaying motion of people walking. Chance correlation of the pedestrians' steps may have caused slight sideways movements of the bridge. It then became more comfortable for the pedestrians to walk in synchronisation with the bridge vibration. This feedback caused the force from the pedestrians to increase the movement of the bridge.

We carried out tests with pedestrians walking over moving surfaces at Imperial College, London, and at the Institute of Sound and Vibration Research at Southampton University. These gave useful preliminary information but were limited in their ability to replicate crowds on a large bridge. The best way to gain data about the events that occurred on opening day was to simulate them on the bridge itself. Therefore, in July 2000, the vibration was measured as a crowd of 100 Arup staff walked on the bridge.

Pat Dallard organised these tests and devised a way of interpreting and analysing the vast quantity of data obtained from them. The results showed that the instinctive behaviour of the pedestrians ensured that the sideways forces that they exerted matched the resonant frequency of the bridge, and this increased the motion of the bridge

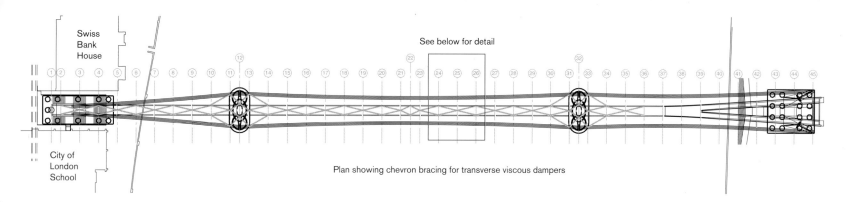

Swiss Bank House

City of London School

See below for detail

Plan showing chevron bracing for transverse viscous dampers

Solution development

Two main ways of reducing the movement were investigated in detail.

Stiffening

The first option was to stiffen the bridge so that the frequencies of the pedestrians' steps did not match that of the structure. Initially this seemed attractive but, as was found during the original design phase, the bridge was already quite stiff from the tension in the cables. The amount of additional structure that would need to be added to increase this existing stiffness would weigh more than the current structure of the bridge.

This option would have involved building a new bridge around and through the existing one, with extra cables on each side of the current set, and a large diagonal truss beneath the bridge deck. Installation would have been time-consuming and costly, and would have dramatically altered the profile of the structure. Moreover, lateral stiffening alone would have left the bridge susceptible to torsional movements, which would in any case have required the addition of dampers.

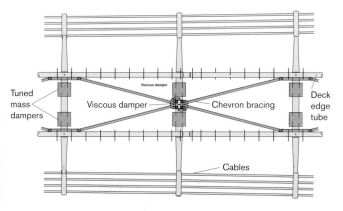

Typical plan showing dampers with deck removed

Tuned mass dampers
Viscous damper
Chevron bracing
Deck edge tube
Cables

Horizontal damping

The second option was to provide damping on the bridge to control all movements.

Horizontal damping is provided primarily by viscous dampers which work in a similar way to car shock absorbers to reduce movement by absorbing energy. A total of 37 viscous dampers have been installed. The majority are situated beneath the bridge deck, on top of the transverse steel arms every 16 metres. Each end of the viscous damper is connected to the apex of a steel V brace, known as a chevron. The apex of the chevron is supported on roller bearings that provide vertical support but allow sliding in all directions. The other ends of the chevron are fixed to the neighbouring transverse arms. In this way the lateral modal movement over 16m is mobilised at each damper. This means that the strokes of the dampers are greater that if for example they were connected directly as diagonals over an 8m bay.

Where possible the viscous dampers have been connected to fixed points such as the piers and the ground. This makes them considerably more efficient since the movement is transferred directly into the damper rather than via a structure which is itself subject to deformation. There are viscous dampers in the plane between the cables and the deck at the piers. These provide vertical and horizontal damping.

In addition, a pair of dampers are located at each side of the approach ramp on the south abutment. These provide damping primarily for the lateral and lateral-torsional modes on the south span. The relationship discovered in our research between the force exerted by the pedestrians and the movement of the bridge means that the greatest damping requirement is for the lowest bridge frequency, which is found on the central span. Four pairs of lateral tuned mass dampers have therefore been added to the central span. Each of these dampers comprises a mass of 2.5 tonnes hung by a double pendulum at each corner. A steel paddle is attached to the mass and sits in a pot of viscous fluid which itself is attached to the transverse arm of the bridge. Viscous damping is therefore generated when the mass moves relative to the transverse arms.

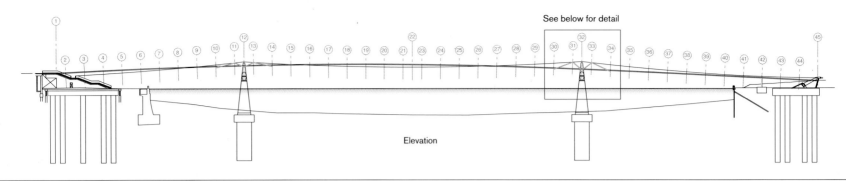

Elevation

Vertical damping

We found no evidence of synchronous vertical excitation on the Millennium Bridge but other researchers report the same effect elsewhere and we have therefore provided vertical damping as a measure of risk reduction. Vertical damping is provided primarily by vertical tuned mass dampers. The vertical tuned mass dampers are arranged in pairs over the traverse arms along the length of the bridge. Each damper weighs between 1 and 2.5 tonnes and is supported on compression springs. As with the lateral tuned mass dampers, viscous damping is provided by the presence of a paddle connected to the mass and suspended in a pot of viscous fluid connected to the transverse arm.

The principles governing these dampers are straightforward, but analysis of their effects on the bridge, once installed, was complex. Mike Willford worked with his team of dynamics specialists inside Arup to find a way of predicting the effect of all of the possible variables in damper performance. A solution involving a series of these viscous dampers was completed in September 2000.

Prototype test

The damping solution was based on data collected in the July crowd test. However we did not want to wait until the next time the bridge opened to find out whether our analysis was correct. Tony Fitzpatrick proposed that we install a fraction of the damping and check how many people were needed to make the bridge wobble.

The test was carried out in December 2000 and the results matched our predictions. Two hundred and seventy-five Arup staff walked up and down the bridge for the day and the viscous dampers and bridge movements were in accordance with the calculations.

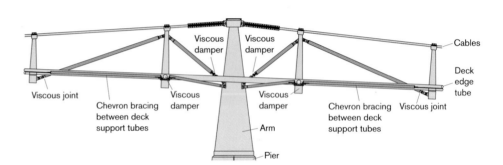

Elevation showing new structural members between arms supporting cables at the south pier

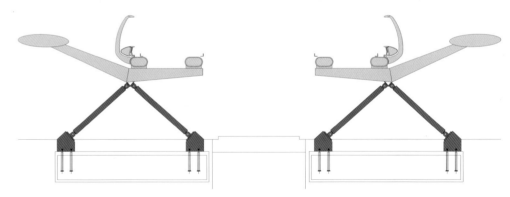

**View from the Bridge,
May 2000.**
Foster and Partners/Nigel Young

Installation of modifications

The dampers were added to the bridge structure through the summer of 2001. Cleveland Bridge UK was the contractor in charge of the operation; Taylor Devices provided the viscous dampers; and Gerb Schwingungsisolierungen provided the tuned mass dampers.

A watershed in bridge design

'Look before you leap' say the critics. 'They pushed technology to the limits and unexpected events like these can sometimes result,' say both supporters and opponents of the bridge.

Both of these statements miss the most important point about the events of Opening Day, which is that the movement that occurred was not related to the structural form of the bridge. If the conclusion drawn by engineers is that they will avoid this problem simply by refraining from designing structures that look like the Millennium Bridge, they risk encountering the same problems.

In the year following the closure of the bridge, we received correspondence which has led us to identify several other instances of this type of movement. These have been bridges of highly conventional construction with lots of visible support. We have also managed to obtain a video of one – the Auckland Harbour Road Bridge which swayed during a large procession in 1975. These examples, coupled with our experimental data, have led us to conclude that any bridge with low lateral frequencies and loaded by a large number of pedestrians is potentially susceptible to the same movement.

Our priority now that the bridge is finally open is to disseminate the information we have gathered as widely as possible. This will change the way all future bridges are designed for pedestrians.

David Newland

**Head of the Engineering Department
at Cambridge University**

Vibration: problem and solution

Fixing the Millennium Bridge's vibration problem has been time-consuming and costly. It's taken well over a year and many millions of pounds to solve a problem that some say should never have occurred. But, whatever your view, the dynamics of vibration are complex and there are many factors involved.

We are surrounded by systems that vibrate: piano strings making music, the heavy drum beat of a disco, a flag waving in the breeze, the wing of an aircraft bending in flight, the rhythmical movement of a train, a car bouncing over a rough road, a ship pitching in rough seas. And there are many less familiar situations. The strings of a tennis racket vibrate after a ball has been hit, a motor bike weaves at high speed when its front wheel wobbles, wind can make a tall chimney shake and electric power lines gallop, and occasionally a bridge vibrates.

A movie film of the spectacular failure of the Tacoma Narrows bridge in the USA in 1940 is often shown to engineering students. Only 4 months after it opened, the world's third longest suspension bridge, with a span more than half a mile in length and weighing tens of thousands of tons, collapsed due to wind excitation. It had been designed to withstand gales of 120 mph but collapsed in a wind of just 42 mph. The bridge deck oscillated so much that eventually the whole structure fell apart.

When I was told that the Millennium Bridge was wobbling, my first thought was that the calculations and testing for wind excitation must somehow be wrong. The Millennium Bridge is very light in comparison with traditional designs and wind could certainly excite oscillations if the aerodynamics were wrong. But it turned out that there was a different cause. With so many people walking across, they were shaking the bridge enough for vibration to build up. Fortunately this movement caused people to stop walking and hold on, and when they did so, the bridge stopped vibrating. The Tacoma Narrows bridge collapsed because the wind that was shaking it didn't stop blowing. When the Millennium Bridge moved too much, people stopped walking so there was no danger of the bridge collapsing. But there was a danger of people falling over and hurting themselves. Something had to be done and Roger Ridsdill Smith has explained what changes have been made since those first unfortunate experiences in June 2000.

How do people or, for that matter, how does steady and usually gentle wind cause a large structure to shake? The Tacoma Narrows bridge was not buffeted by a storm or affected by a swirling or unsteady wind. It responded to a steady cross wind which you would expect only to push steadily against the bridge.

There are a number of factors. The first and most important is that there can only be vibration if the structure is a resonant one. This means that it has a predisposition to vibrate at a characteristic 'resonant' or 'natural' frequency. For example, when the A strings of a piano are played, the predominant tone of their note has a natural frequency of 440 Hz. The strings vibrate through 440 complete excursions every second. This note is determined by the length and tension of the strings and by their weight. But if the same strings are carefully plucked at their quarter span point and simultaneously, in the opposite direction, at their three-quarter span point, a higher note of 880 Hz can be heard. The lower note is called the fundamental and the higher its second harmonic. And there are other harmonics which are higher still and which can also be excited if the strings are plucked in the right way.

Shots from a film showing the Tacoma Narrows Bridge in the USA approaching collapse caused by wind excitation on 7 November 1940. At the time it was the third longest suspension bridge in the world with a centre span of 850 metres
Bettmann/CORBIS

Yamaha C5E grand piano showing
its hammers and dampers.
© *Michael Powell*

The Millennium Bridge is rather similar. Its cables are stretched tight across the Thames with fixed anchorages on the river banks and two fixed points at the supporting piers. Each of the north, south and centre sections has its own natural frequencies. The lowest frequency for the centre (longest) span is about 0.5 Hz, which means that a complete excursion from one side to the other and back again takes 2 seconds as only 0.5 of a cycle is completed in one second. The second harmonic is at about 1.0 Hz, and there is a third harmonic at 1.5 Hz, and so on upwards. It doesn't make much difference whether the cables are vibrating sideways or up and down: their natural frequencies are practically the same.

What about other bridges? They all have their own set of natural frequencies at which vibration can, in theory, occur. Even the massive Forth Bridge has its natural frequencies. Why does vibration build up only on some bridges and not others? Does weight matter, because the Millennium Bridge is a light bridge? Weight helps to prevent vibration, but the Tacoma Narrows bridge was a heavy structure and wind could still excite it. This is because there are other important factors involved.

When a piano is played with its sustaining pedal depressed, each note goes on sounding for several seconds after it has been played. When the sustaining pedal is not pressed, fabric dampers lie on the strings and then notes die away rapidly after being played. Pressure from the dampers dissipates energy which kills the vibration and the note stops. Most bridges, including the Millennium Bridge perform like a gigantic piano string with the sustaining pedal depressed (and therefore the dampers lifted off). Like piano strings, the cables of the Millennium Bridge have a fundamental frequency of vibration and higher harmonics at which they can vibrate with very little damping.

Because they are lightly damped, bridges are therefore capable of sustained vibration if they are excited sufficiently. In the case of a piano, when a note is played, a single large force is rapidly applied to the piano strings and then just as rapidly removed. The piano's mechanism gives the strings a series of hammer blows as music is played. In order to prevent an unpleasant jangling effect, each wooden hammer head is covered by relatively soft felt, but a short sharp impulse is nevertheless applied. This is capable of putting the strings into substantial vibration, which is amplified by the sound board of the piano, and can easily fill a large concert hall with music. But a piano differs from a bridge in this respect. Bridges are not excited by hammer blows. Wind excitation is much more gentle and longer lasting. For there to be a vibration problem, the bridge has to be excited at one of its natural frequencies. How does a steady wind, like the one that destroyed the Tacoma Narrows bridge, do this?

When wind blows against an obstruction, there is a tendency for pressure to build up more on one side of the obstruction than the other. This encourages a wind eddy to form as the wind swings round to one side. Then the reverse happens. The result is that a train of wind eddies (like the puddles formed by the blades of a rowing boat) form as the wind swings first one way and then the other. If the obstruction can move, the changing wind direction encourages it to do so and it then moves sideways in time with the eddies.

There is a characteristic frequency for eddy formation and shedding which depends on the size and shape of the obstruction. When this eddy-shedding frequency is close to the natural frequency at which the obstruction can move from one side to the other, vibration starts, initially very gently, but gradually building up. That is what happened at Tacoma Narrows. The eddies grew and grew as they were shed at the same frequency as the natural frequency of movement of the bridge deck. Vibration occurred because the original steady wind flow was unstable and ready to interact with the bridge in a process called 'self-excitation'. Eventually motion was so great that the bridge broke.

Why doesn't this happen for all bridges? The answer is that the aerodynamic profile and size of the bridge members has to be such that the frequency of natural eddy formation and shedding is different from the natural frequency of motion of the bridge deck. If there is a sufficient difference between these two frequencies, motion does not build up. If the frequencies are too close (and there is not much damping), it does. So a great deal of care is taken when designing new bridges to calculate these frequencies and then wind tunnel tests are done on models to confirm that the calculations are correct.

These calculations and tests were done for the Millennium Bridge at its design stage and there was nothing wrong here.

A different phenomenon, which at the time was not well known to bridge engineers, was at work. Instead of a steady wind causing oscillations to build up, the regular movement of many people as they crowded across the bridge was causing the bridge to begin to wobble sideways. The wobble built up gradually until people became in danger of falling over and instinctively stopped walking. It was a case of people excitation, not wind excitation. There had been examples of people excitation before, but they had not been well-publicised, and engineering standards made no mention of what happened in June 2000.

It had been known for generations that marching armies should break step when they cross a bridge. This is because the rhythmical tramp of massed marching feet could be at a frequency that would coincide with one of the bridge's natural frequencies for vertical movement. Then the bridge would be forced into resonance and might bounce up and down alarmingly. Some ten years before the idea for the Millennium Bridge was conceived, a long footbridge was built over railway tracks at Cambridge station. Fearing the possible adverse outcome of marching armies of undergraduates, tuned mass vibration dampers were fixed to the bridge to artificially increase its inherent damping. They work in the same way as a damper on piano strings.

Notice to troops to break step before crossing the Albert Bridge in London
Chris Patey

Marching armies of soldiers or undergraduates are examples of forced excitation when regular alternating vertical forces are applied to force a structure to move. The effect of forced excitation is well understood and considered whenever bridges (or grandstands and disco halls) are designed. But the crowd walking over the Millennium Bridge in June 2000 had no intention or expectation of making the bridge wobble sideways.

Unwittingly, the crowd became a source of self-excitation to the bridge. Like wind eddies forming at the same frequency as a natural frequency of the Tacoma Narrows bridge, people began to sway from side-to-side at the same frequency as one of the natural frequencies for sideways motion of the Millennium Bridge. People naturally sway very slightly when they walk. So walking at two paces per second produces a small fluctuating sideways force at 1 Hz, close to one of the bridge's natural frequencies. When people swayed very slightly, the bridge wobbled slightly. This made people sway more, and a vicious circle of events began, which only stopped when people stopped walking and their rhythmical movement was interrupted. Mechanical engineers call what happened a feedback problem, because movement of the bridge feeds back to the walking people. Because there is positive feedback, when the bridge wobbles, people sway more, and motion builds up progressively. Eventually people are forced to stop walking, or worse, they fall over.

Self-excitation is primarily the result of an unfortunate coincidence of the natural frequencies of people swaying as they walk and the bridge deck vibrating sideways. Although the natural sideways movement of a person walking, first to one side and then the other, is very small, if it occurs at a natural frequency of sideways motion of the bridge deck, then it can cause this motion to build up gradually unless there is sufficient reason to stop it doing so. Because people tend to synchronize their pacing frequency to the rhythmical movement of the bridge deck, and to sway more when the bridge moves under them, they unwittingly seek out the bridge's vibration.

Why was there such a dramatic effect? All bridges have natural frequencies of vibration, so there was nothing unusual there. All bridges have relatively low damping, so the Millennium Bridge was not particularly unusual there either. The key factor was the light overall weight of the bridge combined with the large weight of people that it can carry. In engineering terms, the 'live' load was high in relation to the 'dead' load. This meant that more than the usual amount of structural damping was needed to prevent vibration building up. The piano's dampers had to be firmly down to stop the self-excitation process occurring. So the solution has been to increase the bridge's damping. That is the key. The best way to prevent self-excitation causing the Millennium Bridge to wobble is to have enough damping present.

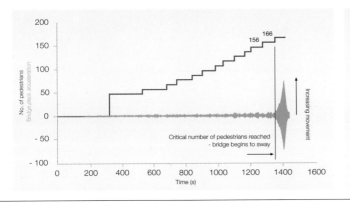

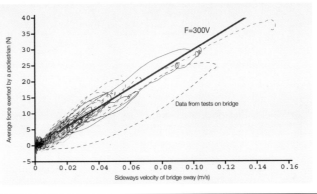

As Roger Ridsdill Smith has explained, much of the damping will be provided by viscous dampers (like car shock absorbers). These are mostly connected within the structure where small relative movements occur, but some are attached to fixed points at the two bridge piers and the south bank. Also there are now a lot of tuned mass vibration absorbers which are additional energy dissipating devices. The resulting structure will have the highest damping, size for size, of any comparable engineering structure. You could say that the original structure has been turned into a dynamic machine. The machine can still move but it will no longer be capable of being excited into large vibrations by an orderly crowd of people trying only to walk steadily across it.

Everyone was taken by surprise by what happened when the bridge was opened. Should the designers have known what to expect? That question has been hotly debated. It's easy to be wise after the event – the intensive publicity means that in future everyone who designs footbridges will know what may happen. Arup deserve congratulations on making available so readily what they have discovered about footbridge design and self-excitation by people. But should they have known more, earlier?

Other people, in Japan, Germany and Switzerland, had had similar experiences but none of these were given the publicity that surrounded the opening of the Millennium Bridge. Hugo Bachmann is a professor in the Swiss Federal Institute of Technology. He had written on a number of occasions about a footbridge that had to be modified because of its large horizontal vibration, and he described observations that pedestrians adjust their step subconsciously to vibration of the deck. And there was an important publication by Professor Fujino and his colleagues at the University of Tokyo which dealt with self-excitation by human walking in more detail. It was published in 1993 in the English language journal *Earthquake Engineering and Structural Dynamics.* This was a curious choice for the paper because it is a journal that is not well-known to bridge engineers. Had the Millennium Bridge's designers known about these publications before, they might well have realised the difficulties that lay ahead.

However, the principal international bridge standards, which bridge designers work to, did not say anything about pedestrian self-excitation (or synchronisation as this has also been called) and Arup conformed faithfully and in all respects with the requirements of all relevant standards.

Like the Tacoma Narrows bridge which was rebuilt as a stiffer structure, the Millennium Bridge could have been stiffened. This would have prevented the coincidence of frequencies that is necessary for self excitation. Stiffening would have raised the bridge's natural frequencies and so taken them above the natural swaying frequency of people when they walk. But stiffening was an expensive option and it would have altered the appearance of the bridge which would no longer be the graceful, free structure that is so admired. The only way to prevent self-excitation of the Millennium Bridge by people was to add the damping devices that are now part of the elegant structure we see today. They are likely to set a new trend in pedestrian bridge design.

Construction
Julian Anderson

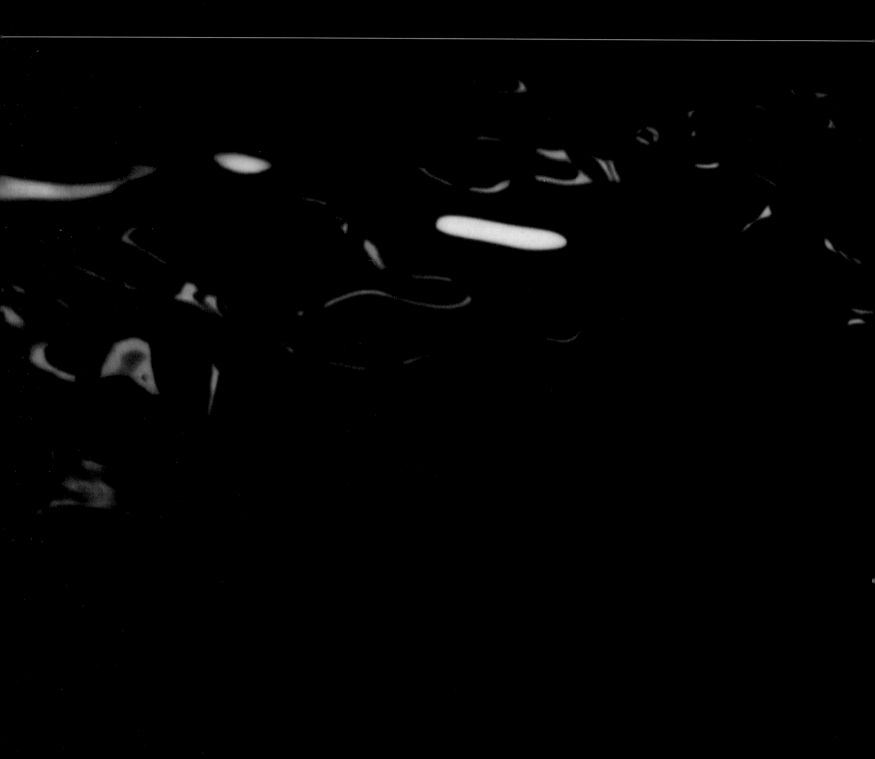

Peter's Hill Steps, March 1998.

Opposite page: Bankside and
the Thames, 27 January 1999.
This page: demolition, clearance
and infill of north site, 20 January
to 9 February 1999.

Preparations for pile-driving,
north site, 12 to 22 February 1999.

Opposite: clearance of
south site, 22 February 1999.
This page: demolition of
coal jetty, 12 February 1999.

**South site archaeological dig,
22 February to 25 May 1999.**

Construction of piles, north and
south sites, 26 May to 25 August 1999.

Concrete pour, north abutment,
14 to 19 October 1999.

Exterior and interior of cofferdams,
16 July to 8 December 1999.

This and following pages:
installation of cable supports,
south abutment, 7 January to
16 February 2000.

**Installation of north pier arms,
11 January 2000.**

Cable fabrication,
12 January 2000.

Sir Anthony Caro inspecting
the fabrication of his sculptures
for Peter's Hill, 11 February 2000.

This and following pages:
installation of highline and cables,
1 to 28 March 2000.

This and following pages:
Delivery and installation of the decks.
April to May 2000.

Views
Dennis Gilbert

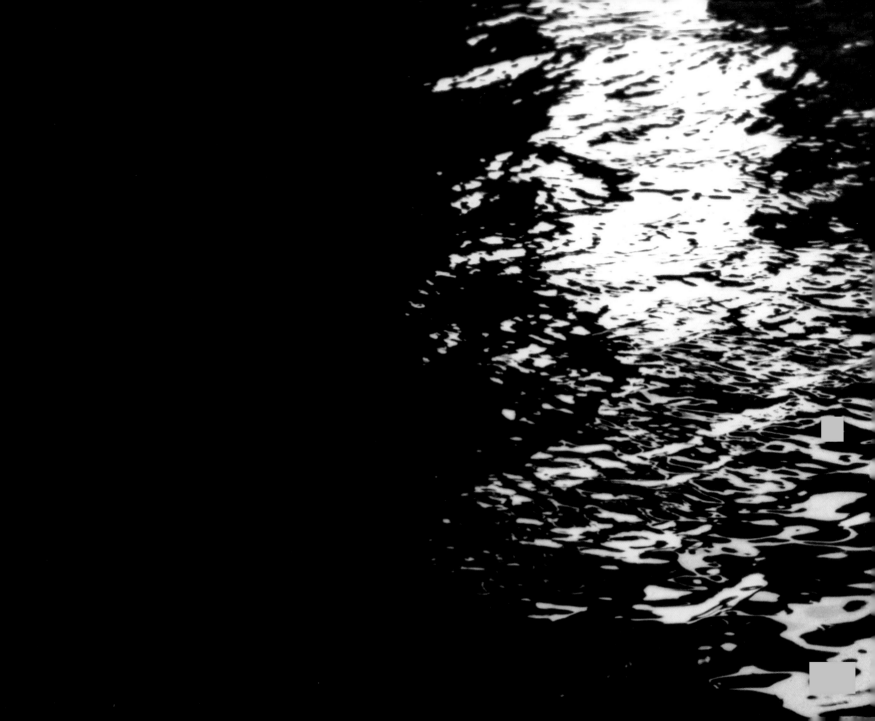

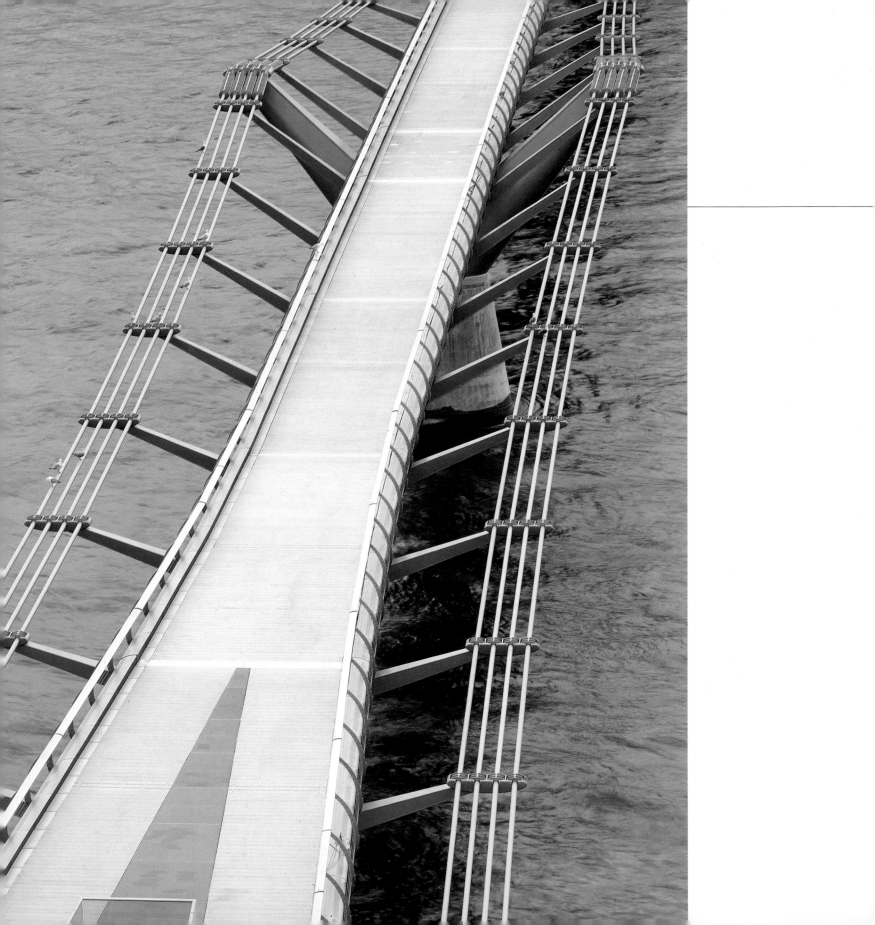

Ian & Linda Coull
Peter & Brenda Curtis
George Anderson Davidson
Delwyn D Dennis
Ed Dicks
The Dormand Family
John & Wumi Dowland
Geoff & Rhoda Down
Roger C Drayton
Francis & Kate Drobniewski
Jill, David, Claire & James Drummie
Bill Duggan Jnr
Shirley Duggan
K R Dunlop, L F & A F Donaldson
Alan Dunsmore
Luke Dunsmore
Rachel Dunsmore
Toby Eckersley
Clive Stanley Efford, MP
Brian H Elton
Michael Entwistle
Ray F Everett
Salman Khalaf Farhan
Shahagir Bakth Faruk
Mansukh Malde Fatania
 & Rita Fatania
James Tsouladze Fernando
Alexia Fetherstonhaugh
Winston & Jean Fletcher
The Flow Foundation
Gemma Fowler & Ben Hamilton
Sheila & Herbert Foxton
The Franey Foundation
Chris Frankland
Valdemar Frantzen
David A Gill
Susannah Maybury Glynn
Michael Godbee
Emily Godwin
Nicholas Goodison
Jennifer Florence Gordon
Jeremy Gould
James Gould
Poppy Gould
Henry & Joanna Greenfield
Alan David Gregory
Anne Griffiths
The GSWD 2nd Charitable Trust
Clifford & Sooozee Gundle
Barbara & Michael Gwinnell
Mrs Sheila Joan Haes

Richard Haines & Janet Wilkinson
Christine & Richard Handley
Richard Hardman
R A & R O Harvey-Kelly
The Harvey-Piper Family
The Hatley Family
John Hauxwell, FIBM
Margot Hayhoe
Norman W Head
Christine Helm
Rachel Charles Kati & Luke
Henderson
Peter & Mary Hinckley
Robert Hiscox
Arabella Godfrey Hobson
R Katharine Holder
F, I, G, G, A, B & A Hudecz
Bill Hunter
Hugo B Hutchins
Joshua J Hutchins
Ray & Margaret Hutchinson
Simon Ingall
The Institution of Civil Engineers
Mary Jo Jacobi
Jane & Scott James
Nigel, Julia, Jamie & Ben Jenkins
Patrick & Jayne Jennings
Alan & Jane Jewell
Lucy Jewell
Tom Jewell
Mark G Johannes
Barbara Jones
John Kean
Fredi & Audrey Keller
Joanna & Richard Kennedy
Ian David Kernohan
Judith King & Andy Poole
Debbie Knight, Deborah Ann Wood
Shirin, Kassamali & Mahomed Ladha
David & Mary Laing
Rebecca Laing
Sophie Laing
Nick Land
Henry Oliver Leaman
Ruth & Stuart Lipton
Eric Lister
Helen Lister
Dr Mel Lobo
Mark & Liza Loveday
David Lynch
Karen Lyons

Bruce MacGillivray
Ruth MacGillivray
La Famille Maher
Hilda & Eric Mallalieu
Rosemary Mallard
Camilla Mason
Oliver Mason
Dr & Mrs A H Masry / Aziz & Khaled
James Bernard McArdle
John T & Jeanne McCombie & Family
Judy & John McCuin
Fiona McKenzie
David Medd
Ewald Meiser
Mr M Meredith Brown, RD FRCS
Kate Middleton
Mark & Judy Moody-Stuart
Peter Moore
Patricia Frances Morgan
Richard N H Morley
Ronnie Morris
Sarah Morris
Faith Nahum
Renate Nahum
Robert M Nelson
Angela Neuberger
Zachary Newman
Graham & Marion Nicholas & Family
Cliff Norris
Mollie & John Julius Norwich
Nic Oatridge
Adam P Oliver-Watkins
Mike & Wendy Olliver
David, Kate & Claire Olliver
Mr Christopher Olumide Olufunwa
The Araba-Olufunwa Family
Ralph Oppenheimer
Marjorie Outlaw
John Overy, BSc
Victor & Oyinkan Oyofo
Sir Peter Parker
Jug Parmar
Chris & Carole Patey
René & Doreen Paul & Family
Robin & Diana Paul
Martin & Pat Perkins
Joan & Michael Perry
Terry & Charles Phillips
Nyda & Oliver Prenn
Sir William Purves
Brian, Jacqui, Alice & Oliver Raggett

Margaret Ramage
 & Michael Karliner
Jonathan Rawes
Guy Reading
B A Reeves
John & Caroline Renard
Viviene Richards
Mrs J & Mr H G Riddlestone, OBE
Sophie & Chloé Ridsdill Smith
Tony Roberts
J A Roberts, Baron Clwyd
Keith & Liz Robinson
John F M Rodwell
Sir Robin Ross
The Ross Goobey Charitable Trust
Richard Rouse
K J Rutherford & D J Lowe
Peggy Salmon
Sally & Anthony Salz
Ro, Martin, Richard & Sarah Samuel
Helga Sands & Julian Darley
The Schimmel Brothers
Sir David Scholey
Jane & Tony Scullion
Allen Paul Seldon
Clare & Michael Shallcross
Andrew & Honor Sharman
Richard D Shaw
Sir Neil Shaw
Wendy Shephard
Joe, Jemma & Rio Sherman
Gilles Shewell & Keith Birch
Mahanta Bahadur Shrestha
Dorothy, Emily & Alfred Simmons
Lynne Slattery
Brian D Smith
Marie H Smith
Samuel & Oliver Smith
Nigel Spalding
Roger Staton
Caroline E J Steenberg
Dennis & Charlotte Stevenson
Honor Stevenson
Joe Stevenson
Tom Stevenson
Rita & Ernest Stickley, MI Mech E
Malcolm Strong
Devika Sukhdeo & Jessica Pearce
John Summers
The Swan Trust
Mr & Mrs Ian Compton Taylor

Peter & Alison Thorne
Shaun W Thorpe
Tony Tresigne
Vanni Treves
The 29th May 1961 Charitable Trust
Hazel & Mike Tyrrell
Michael Uva
Is, Neil, Jenny, Bethan & Ben Vass
Baroness Suzanne von Maltzahn
Edward Walker-Arnott
Andrew M Ward & Susan J Anderton
Bruce, Deborah, James
 & Elliot Warman
The Welton Foundation
George J Westcott
Christopher & Josephine Weston
Jane Whistler
J Charles Whitehead
Matthew Cross Whitney
Paul Leonard Wickham
Derek & Susan Williams
Geoffrey Roy Winch
Bill Wood
Deborah Ann Wood, Debbie Knight
Ben & Christine Wrey
Mr & Mrs E S Zelkha

And in memory of:

Charles E Cannons, M I Struct E
Ronald H Catlin, DFM
Keith Dale
Major C J Davidson, MBE RNO
Joe Gould
Andrew H Harrison, C Eng MICE
F. Marie Louise Jones
Bill Lowther
Marilyn Newman
Florence & Peter Roberts
Leonard Pinfold
Mohanlal Prabhudas Tailor

Millennium Bridge Project Team

Foster and Partners

Foster and Partners is an international studio for architecture, planning and design led by Norman Foster and five partners – Spencer de Grey, David Nelson, Ken Shuttleworth, Graham Phillips and Barry Cooke. The practice's work ranges in scale from Hong Kong's new airport – the largest construction project in the world – to its smallest commission, a range of door furniture.

Representative projects include the new German Parliament in the Reichstag, Berlin, the Great Court for the British Museum, Headquarters for HSBC in Hong Kong and London, Commerzbank Headquarters in Frankfurt, the Metro Bilbao, the Carré d'Art, Nimes, the Sainsbury Centre for Visual Arts in Norwich and the Research Center at Stamford University, California.

Foster and Partners has received more than 220 awards and citations for design excellence, and has won more than 50 national and international design competitions. Norman Foster was Knighted in 1990, appointed by the queen to the Order of Merit in 1997 and in 1999 was honoured with a Life Peerage. He is also a Royal Designer for Industry, a Fellow of the Chartered Society of Designers, an Honorary Fellow of the Royal Academy of Engineering and a Gold Medallist of both the Royal Institute of British Architects and the American Institute of Architects. In 1999 he became the 21st Pritzker Architecture Prize Laureate. The practice has been awarded the Queen's Award for Export.

Ove Arup

Sir Ove Arup (1895-1988) founded his practice in London in 1946. Sir Ove's ideals and principles are as much a driving force today as they were in his lifetime. Foremost among these are the belief in 'total design' – the integration of the design and construction process and the interdependence of all the professions involved, the creative nature of engineering design, the value of innovation and the social purpose of design.

Today, Ove Arup is one of the world's largest and most successful international engineering consultancies, with over 5500 staff working in more than 50 countries through more than 60 permanent offices. Services offered to clients range from planning, feasibility studies, project management and complete engineering design to the supply of specialist skills and advice. Bridges engineered by Ove Arup during the last 25 years include the Øresund Bridge linking Denmark and Sweden, the Hulme Arch Bridge in Manchester, England, and many other bridges in the British Isles, China, France, Ghana, Hong Kong, Nigeria, Sumatra and Turkey.

Sir Anthony Caro

Sir Anthony Caro (b. 1924) is one of Britain's most distinguished sculptors. Over the last 40 years, important museums and galleries around the world have exhibited and collected his work. In 1992, Sir Anthony received the *Praemium Imperiale* for Sculpture, and in 2000 the Order of Merit. He lives and works in London.

Principal Contractors

Monberg & Thorsen McAlpine (Joint Venture), Odense, Denmark, and London, England

Subcontractors

Alumeco Metalhandel A/S, Odense, Denmark (aluminium deck)
Avesta Sheffield, Kolding, Denmark (stainless steel)
Bachy Soletanche (bored piling)
Balfour Beatty, London (enabling works contractor)
Benson Sedgwick Engineering Ltd (sculptures)
Bridon Structural Systems, Doncaster, England (main cables)
William Cook Burton, Burton-upon-Trent, England (tower saddle)
Federal Mogul Sollinger Hütte GmbH, Ilminster, England (bridge bearings)
Gabriel & Company Ltd (architectural metalwork)
Joseph Gallagher (London) Ltd (shaft sinking)
Grotnes Verksted, Rana, Norway (cable clamps)Karl Stahl A/S, Odense, Denmark (stainless-steel wire)
LT Projects (Lightpipe installation)
Masperi Elevatori (inclined lift)
Mostostal, Chojnice, Poland (bridge sections, anchorages – north and south sites)
Mowlem (jetty demolition)
Oxford Geotechnica International (UK) (soil dewatering)
Rautaruukki, Oulu, Finland (steel pylons)
Realtime Civil Engineering Ltd (earthworks, formwork place concrete)
Silver Construction Ltd (pier shutters labour)

Financial Consultants

Davis Langdon and Everest